CONSTRUCTIVISM AND SCIENCE

THE UNIVERSITY OF WESTERN ONTARIO
SERIES IN PHILOSOPHY OF SCIENCE

A SERIES OF BOOKS
IN PHILOSOPHY OF SCIENCE, METHODOLOGY,
EPISTEMOLOGY, LOGIC, HISTORY OF SCIENCE,
AND RELATED FIELDS

VOLUME 44

CONSTRUCTIVISM AND SCIENCE

Essays in recent German philosophy

Edited by

ROBERT E. BUTTS

The University of Western Ontario

and

JAMES ROBERT BROWN

University of Toronto

KLUWER ACADEMIC PUBLISHERS

DORDRECHT / BOSTON / LONDON

Library of Congress Cataloging-in-Publication Data

CIP

Constructivism and science : essays in recent German philosophy / edited by Robert E. Butts and James Robert Brown.
p. cm. — (The University of Western Ontario series in philosophy of science ; v. 44)
Includes bibliographies and index.
ISBN-13: 978-94-010-6921-2 e-ISBN-13: 978-94-009-0959-5
DOI: 10.1007/978-94-009-0959-5
1. Constructivism (Philosophy) 2. Science—Philosophy. 3. Philosophy, German—20th century. I. Butts, Robert E. II. Brown, James Robert. III. Series.
B3183.C66 1989
149—dc20 89-2649

Published by Kluwer Academic Publishers,
P.O. Box 17, 3300 AA Dordrecht, The Netherlands.

Kluwer Academic Publishers incorporates
the publishing programmes of
D. Reidel, Martinus Nijhoff, Dr W. Junk and MTP Press.

Sold and distributed in the U.S.A. and Canada
by Kluwer Academic Publishers,
101 Philip Drive, Norwell, MA 02061, U.S.A.

In all other countries, sold and distributed
by Kluwer Academic Publishers Group,
P.O. Box 322, 3300 AH Dordrecht, The Netherlands.

printed on acid free paper

Softcover reprint of the hardcover 1st edition 1989

Zum Andenken an
Johann Jakob Feinhals
1702–1769
Theologe, Botaniker, Philosoph

TABLE OF CONTENTS

PREFACE ix

ACKNOWLEDGEMENTS xi

INTRODUCTION xiii

SECTION I: CONSTRUCTIVISM AND THE LOGIC OF SCIENCE

K. LORENZ / Science, a Rational Enterprise? 3

C. F. GETHMANN / The Philosophy of Science and Its Logic 19

F. KAMBARTEL / The Pragmatic Understanding of Language and the Argumentative Function of Logic 47

K. LORENZ / Rules versus Theorems 59

J. MITTELSTRASS / On 'Transcendental' 77

SECTION II: CONSTRUCTIVISM AND PROTOSCIENCE

C. THIEL / Philosophy and the Problem of the Foundations of Mathematics 105

P. LORENZEN / Geometry as the Measure-Theoretic A Priori of Physics 127

P. JANICH / The Concept of Mass 145

P. LORENZEN / On the Definition of 'Probability' 163

SECTION III: CONSTRUCTIVISM AND THE VALUE SCIENCES

O. SCHWEMMER / Practical Reason and the Justification of Norms. Fundamental Problems in the Construction of a Theory of Practical Justification 173

C. F. GETHMANN / Protoethics: Towards a Formal Pragmatics of Justificatory Discourse 191

J. MITTELSTRASS / Interests 221

F. KAMBARTEL / Is Rational Economics as an Empirical-Quantitative Science Possible? 241

P. JANICH / Determination by Reality or Construction of Reality? 257

NOTES ON THE CONTRIBUTORS 271

INDEX 279

PREFACE

The idea to produce the current volume was conceived by Jürgen Mittelstrass and Robert E. Butts in 1978. Idealist philosophers are wrong about one thing: the temporal gap separating idea and reality can be very long indeed — even ten or so years! Problems of timing were joined by personal problems and by the pressure of other professional commitments. Fortunately, James Brown agreed to cooperate in the editing of the volume; the infusion of his usual energy, good judgement and good-natured promptness saved the volume and made its production possible. Despite the delays, the messages of the papers included in the book have not gone stale. An extremely worthwhile exercise in international philosophical cooperation has come to fruition; the German constructivist philosophical position is here represented in papers in English that will make its contemporary importance available to a larger audience.

The editors owe thanks to many persons. All involved in the project owe much to the interest and support of Nicholas Rescher, a friend of the undertaking from the time of its inception. My review of the translations was helped immensely by Andrea Purvis' careful copy editing of the typescript. Most of all, however, we owe gratitude and admiration for the tireless efforts on behalf of this enterprise to Jürgen Mittelstrass. The genius of philosophy at Universität Konstanz, for us he exemplifies what is best in contemporary German philosophers: extraordinarily extensive learning, deep philosophical insight, an unperplexed reverence for what is good in the tradition of German thought, a passion for international cooperation and understanding. My contribution to the Introduction was made easier by good conversations (not to speak of good wine and good food!) shared with Peter Janich in Konstanz, Marburg and Pittsburgh; and by good conversations (not to speak of visits to cafes and walks in the woods!) shared with Fritz Kambartel in Konstanz and Meersburg.

ROBERT E. BUTTS

Robert E. Butts and James Robert Brown (eds.), Constructivism and Science, ix.

ACKNOWLEDGEMENTS

The two papers by Kuno Lorenz both appeared originally in English: "Science, a Rational Enterprise? Some Remarks on the Consequences of Distinguishing Science as a Way of Presentation and Science as a Way of Research", in R. Hilpinen (ed.), *Rationality in Science, Studies in the Foundations of Science and Ethics*, Dordrecht (D. Reidel Publishing Co.), 1980, pages 63—78 (© 1980 by D. Reidel Publishing Company, Dordrecht, Holland); and "Rules versus Theorems. A New Approach for Mediation between Intuitionistic and Two-Valued Logic", in *Journal of Philosophical Logic* **2** (3) (1973), pages 352—369 (© 1973 by D. Reidel Publishing Company, Dordrecht, Holland). Both papers are reprinted with the permission of D. Reidel Publishing Company. Carl Gethmann's "The Philosophy of Science and Its Logic" is his translation of the German article, "Die Logik der Wissenschaftstheorie", which appeared in a volume edited by him, *Theorie des Wissenschaftlichen Argumentierens*, Frankfurt (Suhrkamp Verlag) 1980, pages 15—42. "Protoethics: Towards a Formal Pragmatics of Justificatory Discourse", is a translation of Gethmann's paper in German: "Proto-ethik. Zur formalen Pragmatik von Rechtfertigungsdiskursen", in H. Stachowiak and Th. Ellwein (eds.), *Bedürfnis, Werte und Normen im Wandel*, Vol. I (*Grundlagen Modelle Perspektiven*), München & Paderborn (Ferdinand Schöningh & Wilhelm Fink) 1982, pages 113—143. The translation is included by permission of Verlag Ferdinand Schöningh.

Friedrich Kambartel's "The Pragmatic Understanding of Language and the Argumentative Function of Logic", appeared in English in H. Parret and J. Bouveresse (eds.), *Meaning and Understanding*, Berlin & New York (Walter de Gruyter), 1981, pages 403—410. It appears here by permission of Verlag Walter de Gruyter. "Is Rational Economics as an Empirical-Quantitative Science Possible?" is Kambartel's English version of his "Ist rationale Ökonomie als empirisch-quantitative Wissenschaft möglich?" in J. Mittelstrass (ed.), *Methodenproblem der Wissenschaften vom gesellschaftlichen Handeln*. Frankfurt (Suhrkamp Verlag) 1979, pages 299—319. The translation appears by permission

Robert E. Butts and James Robert Brown (eds.), Constructivism and Science, xi—xii.

of Gabler-Verlag. Jürgen Mittelstrass' "On 'Transcendental'" is an English version of his "Über 'transzendental'", in E. Schaper and W. Vossenkuhl (eds.), *Bedingungen der Möglichkeit: 'Transcendental Arguments' und transzendentales Denken*, Stuttgart (Klett-Cotta) 1984, pages 158—182 (appearing by permission of Verlag Klett-Cotta). His second paper, "Interests", is a shortened English version of his "Über Interessen", in J. Mittelstrass (ed.), *Methodologische Probleme einer normativ-kritischen Gesellschaftstheorie*, Frankfurt (Suhrkamp Verlag) 1975, pages 126—159 (appearing by permission of Suhrkamp Verlag).

The English version, "Philosophy and the Problem of the Foundations of Mathematics", of Christian Thiel's "Das Begründungsproblem der Mathematik und die Philosophie", in F. Kambartel and J. Mittelstrass (eds.), *Zum normativen Fundament der Wissenschaft*, Frankfurt (Athenäum-Verlag) 1973, pages 91—114, appears by permission of Athenäum-Verlag. Oswald Schwemmer holds the rights of publication of his "Praktische Vernunft und Normbegründung. Grundprobleme beim Aufbau einer Theorie praktischer Begründungen", in D. Mith and F. Compagnoni (eds.), *Ethik im Kontext des Glaubens. Probleme Grundsätze Methoden*, Freiburg i. Ue. (Universitätsverlag) & Freiburg i. Br. (Verlag Herder) 1978, pages 138—156, appearing here in English as "Practical Reason and the Justification of Norms. Fundamental Problems in the Construction of a Theory of Practical Justification".

Paul Lorenzen's "Geometry as the Measure-Theoretic A Priori of Physics" is an English version of "Geometrie als messtheoretisches Apriori der Physik", in O. Schwemmer (ed.), *Vernunft, Handlung und Erfahrung. Über die Grundlagen und Zeil der Wissenschaften*, München (Verlag C. H. Beck) 1981, pages 49—63; appearing by permission of Verlag C. H. Beck. His paper, "On the Definition of 'Probability'", is an English version of "Zur Definition von 'Wahrscheinlichkeit'", in P. Lorenzen, *Konstruktive Wissenschaftstheorie*, Frankfurt (Suhrkamp Verlag) 1974, pages 209—218; appearing by permission of Suhrkamp Verlag.

Peter Janich holds rights of publication of his article, "Umweltdeterminiertheit oder Konstruktion der Wirklichkeit", in H. Walter and R. Oeter (eds.), *Ökologie und Entwicklung*, Donauwörth (Ludwig Auer Verlag) 1979, pages 92—101, appearing in an English version in this volume as "Determination by Reality or construction of Reality?". His paper, "The Concept of Mass" was written in English especially for this volume.

INTRODUCTION

Philosophy in Germany today is alive and vigorous. It is also inhomogeneous, with strands running in almost every direction.[1] The long tradition of phenomenology from Husserl and Heidegger continues in the works of Gadamer and other hermeneutical philosophers. The rich tradition in political philosophy stemming from Hegel and Marx is embodied in Habermas and other members of the Frankfurt school. But in Germany and elsewhere there has always been another motivation that stimulates work in philosophy: our interest in attempting to understand science. Here the Germanic tradition is nothing short of glorious, running from Leibniz and Kant to the great philosopher/physicists, Mach and Einstein, and the philosophers, Schlick, Reichenbach, Carnap, and very many others. Today this tradition of close association of philosophy and science lives on in Stegmüller's formalization of once orthodox North American philosophy of science. It takes a controversial turn in the work of Gernot Böhme and others on themes grouped together under the heading "Finalization in Science", expressing a manifesto on behalf of the social organization of scientific and technological progress.[2] It is importantly manifested in the unfortunately neglected work of the contemporary German constructivist philosophers of science. *Wissenschaftstheorie* in Germany is philosophy of science in its broadest sense, including work in the logic and the foundations of scientific theories, the conceptual history of science, the cultural and practical context of science, and normative aspects of both scientific and technological progress.[3] The essays by constructivist philosophers in the present volume deal with all of these areas, with a greater stress on philosophical, rather than on historical problems.

Hugo Dingler (1881—1954), a highly original and independent thinker (and a controversial one as well), is the father of constructivism; Kant is taken to be the grandfather. Dingler appears to have been at once a kind of Kantian, an empiricist, an operationalist, a phenomenologist, an intuitionist, and a formalist; and yet he was really not quite any one of these — a situation that reveals how little one learns from philosophical labels. Dingler was principally concerned with the foun-

Robert E. Butts and James Robert Brown (eds.), Constructivism and Science, xiii—xxv.

dations of the exact sciences; the central objective was to answer the Kantian question: "How is exact science possible?" The basic ideas of mathematics and mechanics, according to Dingler, are not derived from experience in the fashion claimed by the empiricists. Rather they are a kind of operational construction, built out of the operations of everyday life; that is, they stem from *practice*.

Understanding the true source of these basic elements is crucial to the reconstruction of the various sciences. This is to be seen in contrast with any ordinary sort of foundationalism; for example, that of the positivists. A so called basic statement or protocol sentence is not basic at all, according to Dingler, but is the result of a complicated constructive operation based on practical human activity.

A leading contemporary constructivist and founder (together with Wilhelm Kamlah) of the "Erlangen School" of philosophy, Paul Lorenzen, discusses this issue by extending a wonderful simile introduced by Otto Neurath (and much repeated by Quine) that knowledge is like a ship at sea; we can only modify or repair it plank by plank, all the while trying to stay afloat. The foundationalist thinks we can put the ship of knowledge in dry dock periodically for a complete overhaul. Like the coherentists who oppose this outlook, Lorenzen thinks the ship is always at sea. But he wants to extend the simile. If the ship never goes to port then it must have been built from scratch at sea. Our ancestors, it would seem, must have been good swimmers. Constructivism would have us get back into the water, and Lorenzen wants to give us all swimming lessons. In this way, he thinks, we can find a "methodical beginning to thought" (1987, p. 6).

The process of discovering our watery beginnings of thought starts with a systematic clarification of our concepts. "Our task", says Lorenzen, "is to build ourselves a ship in the middle of the ocean. The first planks are the predicates, which we first fashion using distinctions offered us by happenstance" (1987, p. 16). This suggests a systematization or taxonomy of concepts in actual use. For example:

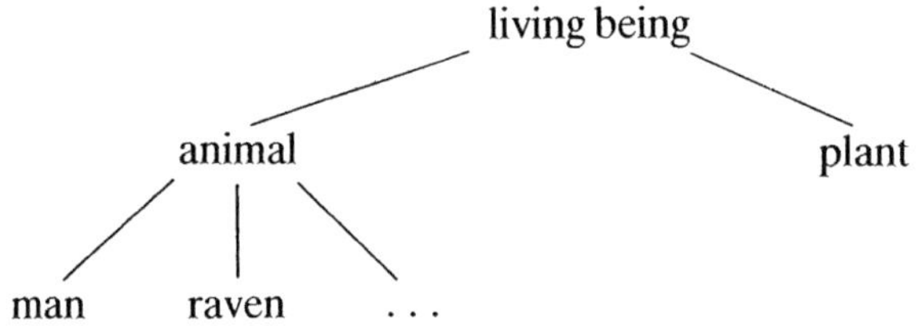

Lorenzen does not call this a taxonomy, but rather a system of "rules" that specify the usage of the predicates. Thus, "*x* being a raven, exemplifies *x* being an animal" is thought of as a rule for the application of those predicates and as a rule that aids in the organization of various predicates into a hierarchy.

That these predicates come to us by "happenstance" is an interesting claim. It suggests that the work of the systematizing constructivist will be done on whatever is the conceptual network of the day, whether that includes "man", "raven", and "living being", or "phlogiston", "caloric", and "vitreous humour". Lorenzen's starting point must then be seen as completely at odds with that of those foundationalists whose first instinct would be to inquire into the legitimacy or the empirical well-foundedness of such concepts. It is also at odds with the initial motivation of philosophical realists who are more interested in discovering the consequences of some conceptual network.

We will postpone for the moment a fuller discussion of constructivism's quest for methodical reconstructions of our concepts. First, it will be interesting to seek to locate the general position of constructivism within the context of positions in other strands of contemporary philosophy of science. We oversimplify to put it so, but it is not too misleading to describe the present situation in Anglo-American philosophy of science (and philosophy of language) as marked by a rivalry of two competing positions. The empiricists take as unproblematic concepts given in experience; these are expressed by the observation terms. All that we can be sure of is described in these terms. Science may employ non-empirical terms such as "electron" or "gene", but the use of such terms is legitimized only by grounding them in the observable realm in one of a number of ways. Some (Mill is a classical example) would say that the very meaning of any sentence containing "electron" is given by an indefinitely long string of observation sentences. Others (Reichenbach, Carnap) link the theoretical to the observational by means of what they called "correspondence rules". More recent empiricists (van Fraassen) think that the question of the meaningfulness of sentences containing terms like "electron" is unimportant; the terms in such sentences are taken to express fictions that are extremely useful in organizing and predicting observations. There are, of course, numerous variations on these themes, and some empiricists, perhaps Mach, would even rule out the use of theoretical terms entirely.

On the other hand, the more realist-minded philosophers (Popper, Putnam, Quine, Sellars) think there is no important distinction to be drawn between the theoretical and the observational languages.[4] They accept that there may be a difference of degree characterizing such languages, but for such philosophers this admission is not of central importance to either the ontology or the epistemology of science. Theories, according to the realist-minded, tend to be understood as conjectures, tested in part by their empirical consequences, but also by various other criteria, some of which may be extra-evidential.

A realist believes that a network of concepts is not derived in any direct way from reality, but is imposed upon nature. The conceptual framework is tried out on nature. The usual result is that nature says "No", and the scientist is then required to go back to the drawing board to design a new conceptual framework to foist upon the world. Empiricists, on the other hand, tend to see concept formation not, as Einstein put it, as "a free creation of the mind", but as something derived from experience, at least when it comes to observational concepts.

It is not possible to locate contemporary German constructivism on either side of these debates: Constructivism must be seen as an alternative to realist rationalism and to empiricism in its present forms. Like the standard realist, constructivists are willing to accept the conceptual network that constitutes present-day science. Unlike the realists, they are not interested in putting it to the test, nor, unlike the empiricists, are they concerned to "reduce" its observational content to a privileged observation language. Rather, they are interested in understanding how this theoretical structure is to be grounded in patterns of human action. They are not so much concerned with questions of hypothesis formation and practices of experimental confirmation within science, insofar as these questions are to be addressed by logical analysis alone. Rather, they are concerned to show how the concepts of science result initially from activity involved in daily practical behaviour. All theoretical concepts are grounded in distinctions made, practical orientations taken, in what Husserl in his later writings called the *Lebenswelt*, the pre-reflexive, pre-scientific, pre-philosophical world that nevertheless guides scientific and philosophical reflection. It is the familiar world in which we all live, a world taken for granted, presupposed — as Mittelstrass will argue, a world that is pragmatically a priori. The model of creation of theoretical concepts is thus human purposive action.[5]

Constructivists believe that application of this model reveals various rule-governed methodical orders whose success is defined by purposes achieved. Scientists do not deal only with sentential theoretical structures, they use instruments and apparatus, they engage in craft activities. Success in building theoretical concepts presupposes skill in the execution of activities of applied craft techniques: Recipes are the final presuppositions of concept formation. Thus constructivists are not faced with the embarrassment of an infinite regress that plagues various forms of foundationalism in philosophy of science. In order to justify some claim *P*, they claim *Q*; in order to justify *Q*, they claim *R*, and so on; in principle, the process never ends. But because constructivists locate the ultimate basis of theory in practical activity, they do not have this problem. Pragmatic operations, unlike propositional truth claims, are the kind of thing that does not stand in need of justification of any sort. A recipe works in application, or it does not. It is the aim of constructivist "reconstruction" to discover what activities and operations successfully generate the various concepts that are central to the sciences. In doing so they will be answering the kantian question: "How is (this particular) science possible?"

It is important to see how crucial for the constructivist programme is this kantian way of putting the philosophical question. For constructivism is a philosophy that seeks presuppositions.[6] It is just that not all presuppositions are like Kant's synthetic a priori judgements or propositions. Mittelstrass (1977) locates a number of other a prioris associated with what Husserl in later works referred to as the *lebensweltliches Apriori*, a priori presuppositions as success stories of pragmatic management of the *Lebenswelt*. He distinguishes between what he calls the *distinction-a priori* (*Unterscheidungs-Apriori*) and the *technical-a priori* (*Herstellungs-Apriori*). Characteristic activities of the *Lebenswelt* involve the making of distinctions and the adoption of orientations. That we draw a distinction between this and that (for example, that we can distinguish red things from green ones) is a kind of brute fact of our *lebensweltliches* experience. We just do note resemblances and differences. On the basis of this distinction making (recall the discussion of Lorenzen above) we generate a simple system of predication together with its logic. This kind of predication yields taxonomies; it is not yet capable of providing mechanisms of proof. In addition, we also find our way about in the *Lebenswelt*; we have a pre-theoretical comprehension of spatial relations and places, which finally grounds, but is not equivalent to, a geometry. Orientation in the world of pre-theoretical experience takes place in what Cassirer called "organic" space and time. It is the space and time of lived activity. In this space I can get about safely, and I can tell you how to get from one place to another only by taking you by the hand and leading you there. It is not a space that I have conceptualized — I cannot draw you a map. The conceptualization of space as something capable of being mapped comes only after I develop instruments for measuring distances. Similar considerations apply to time and mass.

Distinction making and orienting are to be understood as purposive actions. They make arithmetic and geometry and taxonomy (the projection of systems) possible only in the sense that they render intelligible the processes that generate techniques of measurement, which in turn make possible the formulation of powerful hypothetico-deductive empirical theories. I also get about in the *Lebenswelt* by means of simple and again pre-theoretical technical knowledge. I can craft simple tools and instruments. Such activity of making objects becomes rule-governed; we can write down or communicate orally various recipes for proper (methodical) achievement of the ends of craft activities. Taken together, distinction making and orienting and technical know-how or skill constitute the pragmatic a priori pre-suppositions of any kind of eventual theoretical knowledge. Predication and its primitive logic or syntax, rules for the management of moving about in what will later be measured as space and for timing events in what will later be conceptualized as measurable time, and craft recipes all constitute the primitive pragmatic norms on the basis of which we will be able to understand the more abstract concepts of the developed sciences.

The space/time and resemblance/difference orientations and distinctions of the *Lebenswelt* do not, however, comprise the only a priori conditions on theory production. Intermediate between fully formed hypothetico-deductive empirical theories and the *lebensweltliches Apriori* is the *measuring-a priori* (*messtheoretisches Apriori*), which is itself an elaboration of the *Herstellungs-Apriori.* For example, in physics, full theory presupposes the intermediate theory of measurement (measurement of lengths, times and masses). This measuring-a priori is built up on the basis of the pre-theoretical pragmatic a priori. The norms that come to guide our predicating and orienting, plus the craft knowledge involved in elementary technical making of artifacts, eventually provide the grounding for our more sophisticated understanding of measurement. If you like, the intermediate theories of measurement provide the capability of mapping lived space and time onto representations of distinctions made and orientations accepted.

On the basis of this "graded a priori", all grades of which depend ultimately upon purposive practice in the *Lebenswelt*, we finally come to construct hypothetico-deductive empirical theories, and to test and evaluate them in ordinary ways. But in what sense are we to understand that these dependency conditions are a priori? As Mittelstrass emphasizes in his (1985), the three a prioris do not provide *proof structures* on behalf of empirical theories; no proof theoretical a priority is involved. Neither do the pragmatic conditions and the intermediate theory conditions provide *justification* for the theories that depend upon them. Mittelstrass's a prioris are not like Kant's (synthetic a prioris): They do not give epistemological warrant for the production of theories of just certain kinds; they are not conditions of the possibility of experience (in Kant's sense). Rather, the a prioris are "beginnings" or "groundings" of full-blown theory. This concept of grounding (*Begründung*) is fundamental for Mittelstrass and other constructivists.

The appeal to the distinction, technical and measuring conditions as a priori is an appeal of reason. For constructivists, that which is a priori provides part of the basis for taking the enterprise of science to be rational (Mittelstrass is fond of likening grounding conditions to the principle of sufficient reason). Thus to hold that theory depends ultimately upon practical action, upon human purposive behaviour, is to hold that *our understanding* of theoretical science cannot be said to be philosophically complete until we have exhibited (reconstructed) its conceptual content as a constructed consequence of basic human actions. To know how science answers our "Why?" questions presupposes that we can give prior answers put in terms of "I do it because . . .". The a priori conditions on theoretical knowledge are then intelligibility conditions. Put in Kantian terms, they are not conditions of understanding, but conditions of understandability. Thus understood, the rationality of science is radically reconceived, with fundamental status accorded to pragmatic and operational considerations.

Perhaps this will suffice as a general introduction to the novel position in *Wissenschaftstheorie* taken by the constructivists. But it would be a mistake to conclude from our discussion so far that constructivists are only interested in providing methodical and systematic reconstruction of the concepts of the sciences. Constructivism is critical or normative as well, in the sense that by pointing out the different status of various propositions it can have an effect on scientific behaviour. We can illustrate with two popular examples.[7]

Newton's first law says that bodies maintain their velocity unless acted upon by a force. Is this a factual statement? Not according to some constructivists, who point out that the way a force is defined is by way of its changing the velocity of a body. There are no two independent operations that can be carried out, one for velocity and the other for force. The second example concerns temperature. Is the sentence "Water boils at 100 degrees centigrade at standard temperature and pressure" a factual statement? Again, the answer would seem to be "No". Since thermometers are calibrated with this result in mind, any attempt to measure the temperature of boiling water would really just be a test to see if the thermometer had been properly calibrated. This example, like the first, seems to provide us with conventions rather than facts.

The examples point out one of the major normative strategies of the constructivists. Whereas one aim of constructivism is to show which operations correspond to which concepts, the other side of this operationalist coin is to show that we are often naive about the roles played by various elements of our theories, that some concepts do not correspond to any unambiguous operation at all; their role in science is somehow or other conventional.

There is much that is persuasive in this line of thinking. But it should be noted that realists would be inclined to reject the conventionalist conclusion. Consider the example of the thermometer. A response to the conclusion of the constructivist might go like this: It is true that thermometers are calibrated by using boiling water, but there is nevertheless a fact of the matter lying behind this. To see this genuine fact of the matter translate the question to the Fahrenheit scale. Now we can ask: "Does water boil at 212 degrees Fahrenheit?". The "Yes" answer has much less the feel of convention about it. If it is a fact, and not a convention, that water boils at 212 degrees Fahrenheit then the equivalent statement that it boils at 100 degrees centigrade cannot be a convention either (at least in some sense). However, it must be admitted that the *translation* of 212 degrees Fahrenheit to 100 degrees centigrade is itself quite possibly a convention. It is not a clear-cut issue; there is room for more argument here on both sides.[8]

Much of the operational character of constructivism can be seen in Lorenzen's and Lorenz' *dialogic*. Lorenzen's system is similar in result to the logic of the intuitionists as formulated by Heyting. Instead of being a body of truth existing independently of us that all (correct) thought must conform to, dialogic is rather the result of idealized dialogues. The tautologies (the logical truths) of dialogic are the propositions for which a proponent will have a winning strategy. Here, for example, is the dialogical proof of $P \rightarrow (Q \rightarrow P)$

	Proponent	*Opponent*
1.	$P \rightarrow (Q \rightarrow P)$	
2.		P
3.	Why P?	
4.		Proof of P
5.	$Q \perp P$	
6.		Q
7.	Why Q?	
8.		Proof of Q
9.	P	
10.		Why P?
11.	See step 4.	

Anyone who asserts $P \rightarrow Q$ is obliged to prove Q should an opponent come up with a proof of P. As the above idealized dialogue clearly shows the assertion of $P \rightarrow (Q \rightarrow P)$ will always be successful. In the same fashion the familiar logical truths of standard intuitionistic logic will be seen to result from forced winning strategies based on elementary reasoning patterns. The classical law of excluded middle, P or not-P, is not assertable in dialogic; the law of excluded middle cannot be successfully defended in every idealized dialogue, nor can other laws characteristic of classical logic; for example, the law of double negation, $\neg\neg P \rightarrow P$. *Dialogic* does not introduce new proof theoretical assurances on behalf of logical truths; rather, it displays the pragmatic context of discussion that generates or grounds logical truths.

The nature and role of logic is the subject of the essays in this volume by Gethmann (essay 2) who argues for the use of constructive logic in the sciences, by Kambartel (essay 3) who investigates the role of logic in the understanding of language, and by Lorenz (essay 4) in his attempt to mediate between the constructivist outlook and the attitude of the classical logician. The first and fifth essays in the first section of the book (a section largely devoted to methodological issues), by Lorenz and Mittelstrass, respectively, serve as good introductions to the general programme of the constructivists.

As one might expect, the mathematics of the constructivists diverges from that of classical mathematicians whose platonism led them to accept the view that sets, including infinite sets, are independently existing extra-mental objects. The constructivists follow Aristotle and Kant in holding that the only infinities allowed are potential infinities. In certain respects they are carrying on the tradition in the philosophy of mathematics that was so forcefully articulated and promoted early in this century by Brouwer: intuitionistic mathematics. The central idea of Brouwer's intuitionism is that mathematical objects are constructed out of basic mental objects; a proof is a mental construction and as such it is completely self-evident. What gets written up is a report of a mental activity that is essentially nonlinguistic. Such a report is not to be confused with the actual proof, but it will be very useful in helping others to carry out the same mental constructions. That this tradition is being carried on will be evident to readers of Thiel's paper (essay 6) on the foundations of mathematics.

"Protoscience" is the name often used for the constructive study of the foundations of any science. For instance, a distinction is made between physics proper (specifically, mathematical physics), and a systematic study of the distinction-a priori and measuring-a priori concepts that ground physics. The study is known as protophysics, a kind of system intermediate between actual human purposive activities and the recipes for methodical control of such activities and hypothetico-deductive theoretical physics itself. Protophysics is the constructive study of concepts like space, time and mass, all of which are regarded as resulting from methodical generation of techniques of measurement. In this respect protophysics, and more generally protoscience, is an a priori science, linking elements of the pragmatic distinction-a priori domain of the *Lebenswelt* with concepts like space and time through consideration of measuring-a priori rules for determining lengths and intervals of time.[9] The articles by Lorenzen on geometry (essay 7) and by Janich on mass (essay 8) are paradigm examples of some of the important details within constructivist protophysics. A similar approach to the foundations of probability theory is exemplified in Lorenzen's short piece on that subject (essay 9).

Whereas the papers in the first section of this volume deal with methodological concerns, especially logic, and those in the second section deal with the foundations of mathematics and physics, those in the third section are devoted largely to issues in the social sciences. The protosciences include protosociology, protoeconomics, protopsychology and other protosciences not included with the physical sciences and mathematics. Schwemmer (essay 10) and Gethmann (essay 11) are primarily concerned with normative issues; Mittelstrass (essay 12) deals with interests and their role in the social sciences; Kambartel (essay 13) investigates the foundations of economics; and finally, Janich (essay 14) attempts to answer the question "How is ecological developmental psychology possible?".

The essays in this volume thus are seen to cover a large variety of philosophical problems. They provide an interesting introduction to the essential strategies of the constructivist programme, but at the same time they reveal that constructivism allows for differences in emphasis and for experimental accommodations. Constructivism is not a "school" philosophy in the old-fashioned sense of requiring uncritical and unwavering allegiance to a common set of dogmas. Indeed, the essays are worth study even if one forgets that they are written by "constructivists" and learns some philosophy — by whatever name — from them.

NOTES

[1] For a clear and useful critical survey of recent trends in German philosophy see Bubner (1981).

[2] See Schafer (1983).

[3] It is remarkable that Anglo-American philosophy of science, whose practitioners once limited their enterprise to a study of the *logic* of science, has now become broad enough in its concerns to encompass pretty much what is covered by the German *Wissenschaftstheorie*.

[4] Bear in mind that we are presenting a kind of outline of major emphases in contemporary Anglo-American philosophy of science. We are aware of the significant differences that divide a pragmatist like Quine and a scientific realist like Sellars.

[5] Husserl's *Lebenswelt*, as conceptualized by his hermeneutical followers, finally is seen as a world of human discourse. It is that, in part, for the constructivists, for whom philosophy of language is a basic concern. However, the constructivists have a much broader conception of the lifeworld, taking it to include, as we will see, all forms of norm-governed human action. For them, grammar is only one such form.

[6] Our discussion of this theme trades heavily upon Mittelstrass (1977) and Mittelstrass (1985).

[7] Both examples are used by Peter Janich in the final essay in this volume.

[8] Indeed, the authors of this Introduction do not agree on the matter of the "feel" of the "Yes" answer to the question put with respect to the Fahrenheit scale. One author thinks the "feel" is conventional for both measurement scales. This aside, the issue raised here amply confirms the power of the constructivist programme to raise fundamental problems concerning the epistemic status of our scientific concepts. That those problems can arise for non-constructivists as well seems equally true.

[9] Bubner (1981, pp. 150—51) claims that Kant's *Metaphysical Foundations of Natural Science* is a work in what the constructivists now call protophysics. This is only true with certain qualifications. Kant's "special" metaphysics of matter that allows him to "construct" the four momenta of matter (quantity, quality, relation and modality) is indeed a pure a priori physics, but it is also an instantiation of his transcendental philosophy (*Critique of Pure Reason*, B109—110), given that each moment of matter instances a category. Kant's transcendental programme made the "special" metaphysics a part of his general epistemology of science based on the categories and the forms of space and time. As we have seen, constructivism replaces this Kantian appeal to a priori epistemic forms with an appeal to norms of action. Nevertheless, and despite this crucial difference, Kant's *Metaphysical Foundations of Natural Science* does appear to be executed in the spirit of a later constructivist understanding of science. The best work in protoscience available in English is Janich (1985). Present-day Anglo-American philosophers of science will for the most part find protoscience an unwelcome and redundant theory; for most of these philosophers concept formation and the theory of measurement are themselves determined by fully developed scientific theory.

REFERENCES

Bubner, R. (1981), *Modern German Philosophy*, trans. by E. Matthews. Cambridge: Cambridge University Press.

Janich, P. (1985), *Protophysics of Time: Constructive Foundation and History of Time Measurement*, 3d ed. Boston Studies in the Philosophy of Science, vol. 30. Dordrecht: D. Reidel.

Kambartel, F. (1980), "Pragmatic Reconstruction, as Exemplified by an Understanding of Arithmetics", *Communication & Cognition 13*: 173—182.

Lorenzen, P. (1969), *Normative Logic and Ethics*. Zurich: Bibliographisches Institut Mannheim.

Lorenzen, P. (1987a), *Critique of Political and Technical Reason* (The Evert Willem Beth Lectures 1980), *Synthese 71*: 127—218.

Lorenzen, P. (1987b), *Constructive Philosophy*, trans. by K. R. Pavlovic. Amherst: The University of Massachusetts Press.

Mittelstrass, J. (1977), "Changing Concepts of the *a Priori*", in R. E. Butts and J. Hintikka (eds.), *Historical and Philosophical Dimensions of Logic, Methodology and Philosophy of Science*. Dordrecht: D. Reidel.

Mittelstrass, J. (1980), "Towards a Normative Conception of the Growth of Knowledge", *Nature and System 2*: 231—244.

Mittelstrass, J. (1985), "Scientific Rationality and Its Reconstruction", in N. Rescher (ed.), *Reason and Rationality in Natural Science*. Lanham, MD: University Press of America.

Schafer, W. (ed.) (1983), *Finalization in Science*. Boston Studies in the Philosophy of Science, vol. 77. Dordrecht: D. Reidel.

ROBERT E. BUTTS

JAMES ROBERT BROWN

SECTION I

CONSTRUCTIVISM AND THE LOGIC OF SCIENCE

KUNO LORENZ

SCIENCE, A RATIONAL ENTERPRISE?

Some Remarks on the Consequences of Distinguishing Science as a Way of Presentation and Science as a Way of Research

Our main concern has been to deal with various aspects of the concept of rationality. Specifically, the rationality of some scientific activities has been questioned, among them the case where doxastic attitudes are chosen, or where weights are assigned to experts in a group which aims at a consensus on some scientific question, or even where the guiding principles necessary for calling an activity 'scientific' are being adopted. More generally, we asked for criteria of rationality with reference to human behavior in verbal or non-verbal interactions or to human actions (i.e. intentional behavior) in general.

As an immediate consequence of these attempts to reach at an explicit determination of the (meta-rational) norm 'be rational' the question came up whether such a norm should be turned from a categorical one into a hypothetical one, valid only under conditions where rational procedures cannot conflict with what intuitively could be called progress due to imaginative ingenuity, i.e. a kind of "sound" irrationality counterbalancing the tendency towards a non-sensitive conservatism felt to be inherent in any kind of rationality. If, now, the principle hidden behind that progress is the old and venerated quest for truth the question arises whether rationality as a condition on means for truth as an end is really able to serve as a safe guide to truth. In other words, is there a chance to formulate conditions of truth for the results of scientific activities without recourse to the rationality of those activities: is it possible – to use a metaphor of Wittgenstein – to throw away the ladder after having climbed up on it to true results. Or, is the very claim for truth nothing but a reasoned and, hence, rational way to attain results.

In the following remarks I want to make use of the distinction between science as a way of presentation and science as a way of research, which may throw some light on that seeming conflict among the two norms 'be rational' and 'be right'. The idea is that science as a way of presentation should be understood as a theory of meta-competence (the result of following the directive: be rational!) – a knowledge of the means to secure the truth of propositions about objects –

Robert E. Butts and James Robert Brown (eds.), Constructivism and Science, 3—18.

whereas science as a way of research is a theory of object-competence (the result of following the directive: be right!) – a knowledge of the objects through adequate representations of them. Object-competence can, hence, be acquired only in the presence of the respective objects; you have to deal with objects – appropriate speech acts included – to get acquainted with them. Meta-competence, on the other hand, works primarily in the absence of objects: it is just that tool on the level of signs which serves to counterbalance the lack of "knowledge by acquaintance" through substituting "knowledge by description" for it.[1]

What has to be investigated is the interrelation of these "two ways" historically as well as systematically. I would like to connect my considerations with a historical claim concerning the traditional two ways to cope with epistemic scepticism: rationalism and empiricism. Due to a misunderstanding on the part of the empiricist tradition up to the modern analytical philosophy of science, the ways of research have been erroneously treated as descriptions (i.e. on the meta-level) relative to given domains of objects, and due to a dual misunderstanding on the part of the rationalist tradition up to the modern constructive philosophy of science the ways of presentation have in turn been erroneously treated as constructions on the object-level relative to given domains of concepts. For further support of this claim I continue with a kind of historical sketch concerning the fate of rationalism and empiricism.[2] There are two main developments originating from well-known problems of epistemology in both of these philosophical positions which may be characterized in the following way: Out of rationalism emerges transcendentalism to secure a unique set-up of at least the natural sciences, mathematics included. We may call this the *a priori* method to set up mathematics and some fundamental parts of physics. Empiricism, on the other hand, gave way to evolutionarism, some kind of free choice principle to be used for starting, e.g. the sciences or any other human artefact. We may call this the observation method to stick to what is at hand in a given moment. Now, it is common opinion to treat an epistemology of the first kind as the only way out of epistemic scepticism taken seriously, whereas an epistemology of the second kind bounds scepticism by some commonsense relativism which implies dropping any reliance upon science as a substitute for religion concerning matters of fundamental world view.

This frame for dealing with the claims of scepticism hides that difference of presuppositions in the philosophy of science which I referred to as treating science as a way of presentation and treating it as a way of research. Of course, this difference takes up the old person-oriented *ars iudicandi* on the one hand and the matter-oriented *ars inveniendi* on the other hand, and I should add – just as a remark – that this Leibnizian difference of Analysis and Synthesis, as he alternatively calls these two "artes", is more general than Reichenbach's distinction between a context of justification and a context of discovery, since presentation (unlike Reichenbach's justification) as meta-competence is dependent on object-competence (otherwise the linguistic means cannot be safeguarded against loss of meaning), and research (unlike Reichenbach's discovery) as object-competence is dependent on meta-competence (otherwise mutual communication will lose any control of success or failure).[3]

Transcendentalism searches for justifiable presentations (being true theories of certain domains of objects) whereas evolutionarism represents a way of adequate research (being significant encounters with certain kinds of objects), such that in the first case we get well-founded sequences of propositions, yet in the second case a well-determined network of mutually related objects.

At once a further complication arises. To search for sequences of propositions is certainly not a purely linguistic matter. Instead of just giving a construction of certain objects on the language-level the crucial issue is to judge their "relation" to the object-level with the aim of securing their truth. Analogously, to represent a network of objects cannot be done on the object-level alone; representation is bound to rely on linguistic means with the aim of determining the objects by precise descriptions.

In order to avoid erroneous identifications, I have deliberately used here the terms 'search' and 'represent' to refer to activities within science as representation and science as research, respectively. For, certainly, there is second order research concerning presentations (e.g. within what is called 'science of science') and second order presentation concerning research (e.g. within the well-known 'logic of inquiry'), and neither should be identified with what I called 'search' and 'representation', respectively.

Search within presentations is search for *true* presentation, whereas representation within researches is representation of *significant* research. What is at stake from the purely linguistic point of view is the question of how the verification (and falsification) of formulae is interrelated with the signification of terms. Both questions, the justification of propositions and the constitution of objects, have to be answered separately yet dependently on each other and they must not be confounded.

If the question of constitution is falsely treated as belonging to the problem of justification it yields evolutionarism as a brand of radical empiricism.[4] And again, if the question of justification is falsely treated as belonging to the problem of constitution it yields transcendentalism as a brand of radical rationalism. To keep both questions within their proper bounds has a chance of success only if their mutual dependency is treated clearly and distinctly. This means especially to ask for methods to translate theories, including states of theories – considering theory-change – into each other: In what sense may two synchronically or diachronically different theories have the same content – though saying different things of different entities.

The most prominent example of the radical empiricism – the empiricist misunderstanding – which arises when problems of constitution are treated as if they were problems of determination, i.e. as if they concerned investigations into the truth-conditions of assertions about the constituted objects, can be found in the evolutionary pragmatism of C. S. Peirce.[5] With slight modifications only, this is equally true of the "Analytische Wissenschaftstheorie" as it grew out of the logical empiricism of the Vienna Circle.

For example, the usual set-up of formalized theories has never been seriously questioned, i.e. the start with given domains of objects on the one hand and sets of predicates together with suitably chosen axioms about those objects on the other hand. There is a freedom of choice in both respects – e.g. phenomenalistic systems may compete with physicalistic systems, and preference for some set of primitive notions and principles should always be treated as a contingent fact itself – yet there is usually no hint as to how somebody can acquire a position enabling him actually to choose among alternatives. This again remains a historical and thus contingent fact.

In the converse case, the most prominent example for the radical rationalism – the rationalist misunderstanding – which arises when problems of justification are treated as if they were problems of constitution, i.e. as if they concerned investigations into the conditions of possible experience (how objects of experience have to behave in order to be accessible to knowledge) can, of course, be found in the transcendental idealism of I. Kant.[6] It is, therefore, not accidental when proponents of the "Konstruktive Wissenschaftstheorie" use arguments akin to Kantian ones to substantiate the claim that what is called 'protophysics' can serve as an a priori foundation for physics.

I will not go into further historical details now, but rather stress certain features of the discussion between the analytic and the constructive philosophy of science which are relevant for the epistemological issue in the sciences I am concerned with.

For convenience of presentation, I will start with the discussion of a thesis, which Harald Wohlrapp has convincingly defended a few years ago:[7] The analytic philosophy of science on either of its three main stages, Carnap–Stegmüller's empiricism, Popper–Lakatos' rationalism, and Kuhn–Feyerabend's historicism, should – according to Wohlrapp's claim – essentially be understood as concerned with science as a way of research, whereas the constructive philosophy of science of the Erlanger Schule and of related positions, is basically concerned with science as a way of presentation.

As an important consequence, the difference of criteria for what shall be considered as science and, hence, as a rational enterprise can be stated. The criteria of science as research are essentially those of success, derived from the *actual* procedures of working scientists; necessary conditions are, e.g. the use of well-defined predicates, the reliance on the consistency of the set of non-derived sentences, reproducibility of operations *et alii*. On the other hand, the criteria of science as presentation follow conditions of acceptability and are in this sense "foundational"; they derive from *potential* procedures of scientists and can be characterized essentially by two principles: the principle of method (i.e. presentations work stepwise without "jumps" – a kind of completeness claim) and the principle of dialogue (i.e. presentations can be criticized, which means it can be questioned whether they fulfill the first principle).

If one proceeds this way and at the same time argues from a treatment of science in the first case as a fact and in the second case as a norm – how science is and how science should be – a certain kind of simplification occurs and affects the conclusion. The reason being simply that research cannot be described without recourse to decisions of what shall count as relevant and that presentation cannot be issued without using the content of the respective predicates. Therefore, it remains to be investigated whether the specific claims of failure and success Wohlrapp raised with respect to the two sets of criteria he discusses really carry conviction as they stand or whether failure and success rather depend on some further distinctions connected with the difference of research and presentation.

To repeat what I said in the beginning, I would like to claim that inasmuch as questions of constitution are confounded with questions of justification – or, to use the linguistic angle, questions of signification confounded with questions of verification – the criteria for science either as research or as presentation will lead to difficulties and eventually to failures. Success, therefore, in either case is dependent on a clarification of the interrelationship between answering "what there is" and answering "what is true", the ontological and the epistemological version of the question of how the two levels of objects and of signs ("world" and "language") separate within and unite into one domain of (scientific) language-games.

I have tried to show elsewhere[8] that this domain has to be understood as a domain of "preactions", beyond the action-act dualism as the prototype of the type-token division, and equally beyond the classical distinction of something given and something to do. The idea is simply to start with – from a later point of view, complex – objects which do not yet bear the differentiation between actor and action or between action and object of an action or result of an action. From that starting point to develop both ordinary language and the language of science as something on top of a more elementary and obviously fictitious language where only reference to such objects occurs, is already a piece of work during which most of what is treated later on as logical or ontological presuppositions of a language gets decided.

The difficulty is that in giving a description of this (re)construction the language of description, i.e. some standard natural language in use, is far more developed syntactically and semantically than the described language during the process of its construction.

Hence, in order to get an adequate account of the construction it is necessary to introduce certain devices which make sure that the description at any stage is not dependent on those features of the syntactic and semantic structure of the language of description which do not yet belong to the structure of the constructed language. For example, the difference of singular and general terms within the language of description should not be relevant for describing the initial stage of construction, where within a fictitious elementary language only reference to "preactions" occurs. Rather, there should exist an explicit step of introducing that very difference within the elementary language. And this is done by proposing to introduce the singular/general distinction on the elementary level of non-analysed actions as the distinction of *schema* and *actualisation*. These descriptive terms refer to the difference of "once", "once more", "once more again", . . . , which is practically acquired in situations of repetitive imitation (=imitative repetition) with respect to any preaction.

It is obvious that such preactions are the pragmatic version of Strawson's "feature universals" in his essay *Individuals* (London 1959), at least with respect to their general aspect; Strawson forgets to include in his presentation their singular aspect as something on a par with the general aspect of preactions.[9]

Now, linguistic signs are the means which have developed gradually through our evolution to articulate the mutual dependency of schema and actualisation with respect to any preaction: it becomes possible to *say* which general object belongs to which singular object, i.e. under which concept a certain case falls or by which case a certain concept is fulfilled. Through language something singular acts as a *symbol* of something general, and, the other way round: through language something general acts as an *aspect* of something singular.[10]

Then, it is perhaps not any more offensive – by being liable to the pitfalls of a remake of the cartesian dualism – to say: the singular gives the empirical base, the general the rational design; the two cannot be separated; research starts with singularia, presentations with generalia. Since theories of both areas exist, hierarchies of theoreticity appear[11] and the situation, including the empirical/rational distinction, becomes confused.[12] Just as a remark, it is interesting to note that these preactions are the candidates which show the two aspects, the mental one and the physical one, in recent discussions about the theory of actions[13] only after the preactions have been developed into proper actions.

As I already mentioned in connection with Wohlrapp's paper, it creates still further confusion, if in the first case the methodological position of the analytic philosophy of science is characterized as being descriptive, whereas in the second case the methodological position of the constructive philosophy of science bears the label of being normative. Inasmuch as questions of constitution have consequences in terms of stipulations concerning the objects of scientific discussions the "definition" (I prefer the more general term 'introduction') of basic predicates about them is included – the insistence on the normative character of some fundamental part of science – let us call it 'proto-science' – is reasonably supported.

Similarly, questions of justification, concerning the context of these objects – the *use* of predicates, so to speak, not their *introduction* – give rise to descriptive aspects of any science.

Here again, it might be useful to recall that in general predicates within scientific languages are defined on given domains of objects – extensionally as certain classes of those objects. Hence, they cannot be treated as primary predicates, they are derivative with respect to the defining predicate for the domain of objects (i.e. the "substances" defined as the instantiations of that primary predicate). We, then, speak of properties, and the usual problems concern questions of whether properties hold of objects which obey certain other descriptions, and they never concern questions of elementary constitution.

Those constitutional questions – unless they are non-elementary, i.e. of second order, yielding domains of abstract objects (such processes are of course well known and extensively treated everywhere) – occur on a language level which is itself of a theoretical nature only. It is the elementary level I referred to earlier and which can now be characterized as the one where terms are introduced, not the one where they are used as in ordinary non-scientific and usual scientific language. E.g. when you assert 'this leaf turns yellow', the constitution of objects like leaves (i.e. the introduction of the term 'leaf') is presupposed, whereas the constitution of objects like "yellows" (nominal use of the term!) is pushed to a second order level: Yellow is constructed as an abstract object, a "quality", turning the word 'yellow' from a non-primary general term – standing for a "characterizing universal" in the sense of Strawson[14] – into a singular term, a nominator, as I propose to call it. What remains in the case of the assertion in question is to judge upon the use of the terms 'leaf' and 'yellow' (or 'turn yellow') which by all standards is a question of true description.

Constructions remain within one language- (or object-)level, descriptions concern two consecutive levels. This difference is well-known, e.g. in logical theory, where formulae can either be constructed by formation-rules, or they can be described by means of suitably chosen predicates of a metalanguage (used for formalising the construction).

The fundamental notions are *partition* (of a whole into parts) in the case of constructions, and *attribution* (of a property to an object) in the case of descriptions. Mereology and set-theory provide the respective formalisations of these notions though their interrelation has by no means been sufficiently clarified up to now.[15]

I hope these remarks give sufficient support for the claim that there is no simple correlation between science as research and stating what is (being the result of research) on the one hand, and between science as presentation and issuing what shall be (being the guarantee of presentation) on the other hand.

In either case the set-up of science is not only a question of justifying a corpus of sentences (used as constatives and/or as directives), but a question of introducing meaningful terms viewed from different angles or better: distances, only, as object-competence from nearby and as meta-competence from far away. Justification now includes the search for the truth (science as theory) together with the search for the good (science as praxis) as much as meaning exhibits both aspects of (theoretical) signification and aspects of (practical) relevance.

According to John Rawls,[16] the good is taken over by justice, though the criticism of Tranøy shows[17] that the good is justice only in its social aspect; in its individual aspect it will have to be determined as freedom – socially granted through mercy according to Tranøy –, and there is no easy conciliation between these two aspects.

The corresponding two aspects of truth look somewhat different though well-known under the labels of truth as consensus and truth as correspondence. In its genuine epistemological context 'truth' refers to a qualified consensus among persons on some matter of common concern, e.g. in the way Peirce had indicated by equating truth with the ultimate opinion of the indefinite community of investigators in the long run.[18] In a more recent terminology it is said that such investigations, if they obey the qualifications (i.e. if they are conducted "rationally"), maximize *epistemic utility*.

Now, the usual correspondence theory of truth as the alternative aspect to truth as (rational) consensus should rather be understood as a treatment of truth in an ontological context: truth gets equated with adequacy of linguistic representations of objects which, again in a more recent terminology, means to conduct investigations with the aim of optimizing *truth output*. We know reasonably well how to handle these two aspects of truth,[19] yet in the case of the good, i.e. the call for procedures to tackle the problem of justifying norms, we are on far less secure grounds. To comment on our discussion of possible conditions for the rationality of actions I should like to stress that the so called means-ends rationality[20] takes care only of those cases where the means-ends distinction is relevant for the action in question. Outside the area of technical (or instrumental) norms – norms count here as generalized directives towards an action type – neither practical (or social) norms, where the required action is an "end in itself"[21] nor moral (or meta-ethical) norms, where rather attitudes towards actions (including forbearances) than actions themselves are of concern, can be treated in this way.

The antagonism of freedom and justice, the individual and the social aspect of the good, we have been observing is due, I claim, to that antagonism of social norms and moral norms which derives from a conflict between two second-order ends: social norms aim at uniformity, moral norms protect diversity. The reason why instrumental norms which aim at uniformity, too, do not essentially, i.e. irrevocably interfere with moral norms lies in the fact that instrumental norms allow substitution of means for the same end.

Now, since language-norms, i.e. the use of a language, whether enforced by institutions or not, seem to me in their conceptual aspect to be cases of technical norms (here: social conventions), yet in its perceptual aspect (when signs are taken as objects in their own right) to be cases of moral norms (here: individual rights to choose one's own way to speak, e.g. the socio-historic background of the language-norms as conventions), one could at once understand why it makes sense to plead for a unified language for science and not e.g. for the arts, and why there is so much opposition to such a pledge since almost every scientific enterprise includes distinctive features of artistic idiosyncrasy.

This may furthermore be taken as a hint that it is again the difference of object-competence and meta-competence which not only governs the two ways of science as research and as presentation, but also two ways to treat the rules on signs, especially language-norms, one perceptually – the way of art – and the other conceptually – the way of science (of course not restricted to the natural sciences).

This brings me back to the main topic, how the difference and the interdependence between science as research and science as presentation can actually be characterized.

The relevant difference of research and presentation as against the simplified accounts I discussed above comes in when we look for the *support* of a scientific theory. As long as science is treated as a way of research, this support should derive *only* from the descriptive power of the theory relative to the singular objects (= perceptual cores) of research, though usually, in the analytic philosophy of science, due to the aforementioned lack of a clear separation between constitution and justification, this support is extended to include the explanatory power of the theory as well. It is common to use the term 'confirmation' (referring to non-elementary propositions in relation to relevant singular "data") in this respect, and this entails a confrontation with the – I dare say unsolvable – riddles and paradoxes of induction.[22] The simple reason for the claim that nothing beyond the descriptive power of a theory can be treated within science as research only, derives from the following considerations: The explanatory power of a theory refers to the kind of interrelations which obtain among the different propositions of the theory, especially to an assessment of the range of validity of fundamental principles like those of conservation in physics. Hence, an account of the explanatory power can be given only by judging upon the conceptual frame of the theory used for the argumentations in science as presentations.[23]

The argument in the alternative case runs conversely: as far as science is treated as a way of presentation, the support of a scientific theory should derive *only* from its constructive power relative to the general objects (= conceptual frames) of presentation, though, usually, in the constructive philosophy of science, due to the same confusion of constitution with justification, this support is extended to the regulatory (and, hence, normative) power of the theory as well. The term 'approximation' (referring to elementary objects in relation to relevant general "idea[l]s") is in this connection occasionally used; and, as an equally disturbing consequence, it becomes necessary to handle the vexations of the is-ought gap.[24] Here again, it is easy to see that in science as presentation any attempt to go beyond the constructive power of a theory and to judge upon its power to issue what shall be – unless this is treated as a second-order question only, i.e. as a question of what kind of scientific activity (rather than objects of activity) should exist – will need reference to the perceptual cores of science as research, e.g. to the encounters with singular objects in experimental situations.

What I should like to claim is that both the explanatory and the regulatory (or normative) power of a scientific theory can be assessed properly and without bias only if the set-up of science cuts straight through the separating line of research and presentation. This means especially that on each level within the hierarchy of theories the link between constitutional and justificational questions – and that refers to the interdependence of constructive with descriptive procedures as well – must not be lost sight of. E.g. the constative metapredicate 'state' and the directive metapredicate 'bring about' on the kernel-sentence 'this is *P*' may be used to arrive at the linguistic representation of two speech-acts, a constative '[I] state that this is *P*' and the directive 'bring about that this is *P*!' such that 'this will be *P*' can be explained by the directive (i.e. a demonstration of a future state of affairs by means of a present volition) and that 'do *P*' can be normatively justified (= regulated) by the constative (i.e. a probation of a present imperative by means of a continuing ability). This shows at least by way of indication how the interrelations in question might be dealt with.

Hence, trying to determine an adequate meaning of the two central concepts (scientific) *explanation* and (scientific) *regulation* amounts to nothing less than a reassessment of whether and how a unified treatment of science is possible. For this purpose, the concept of unified science should no longer be understood in the original historical setting along with a developing analytic philosophy of science. In the light of the considerations just offered I claim that a unified approach to science, unless it falls victim to typical "Scheinprobleme" as the one concerning the possibility of induction or the one concerning a bridge over the is-ought-gap, has to consider procedures with respect to activities both of research and of presentation. It has to develop a concept of science starting with a kind of unity of research and presentation, where the domain of (scientific) language-games uniting "world" and "language" in the sense I have outlined above becomes the result of the first step. These language-games of preactions together with their articulations can then be treated in both of their aspects: matter-oriented (research, unfolding object-competence by introducing acquaintance) and person-oriented (presentation, unfolding meta-competence by using description).[25]

For visualization of what I am driving at, I may use an example of current dispute: the different approaches to (physical) geometry. Concentrating on the research aspect of physics, the (temporal) behavior of (physical) bodies relative to their spatial coordinates is judged with respect to quite general hypotheses concerning space-time-structure (explanation-bias!). The presentation aspect of physics, in the protophysics of the Erlanger Schule on the other hand, asks for a series of steps to introduce the fundamental concepts of geometry, chronometry, and hylometry in that order using "idealized" operations with (physical) bodies (regulation-bias!).

In the second case, what is done is to provide meaningful terms – that they can be used successfully outside presentational questions is taken for granted. It is not surprising that certain propositions come out true a priori. In the first case, something completely different happens: here, propositions about given objects are tested to secure their validity, which means treating them as empirically based. The introduction of the terms used is taken for granted inasmuch as presentational questions are considered to be a *cura posterior*. Though theoretical activities of supplying true descriptions govern science in its research aspect or, rather, because of them, the presentational necessities like introducing meaningful terms get neglected. And, conversely, the concern with practical operations to get adequate constructions of fundamental concepts for science in its presentation aspect seduces into one minimizing the importance of the problem of how those concepts can effectively be used in research situations.

The real issue actually boils down to the question of whether the introduction of meaningful (geometric) terms like 'straight', '*n*-times the length of' etc. can be treated as an *extension* of ordinary language about ordinary objects, serving better criteria of relevance according to further developed standards of significance and truth. For, if extendability fails, we are stuck in conceptual frames without prospects to satisfy them; and if presuppositions serve as substitutes for explicit introductions, there is no chance to guarantee anything beyond the perceptual cores.

It is easy to see that the last two conditional sentences may serve as a modern and more refined version of Kant's famous dictum that concepts without intuitions are empty and intuitions without concepts are blind. The refinement consists in the introduction of hierarchies of theoreticity starting with (of course not uniquely determined) common sense experience phrased in everyday language.[26] The domains of objects of scientific discourse have to be arranged on levels of ascending and descending order without any chance to argue definitely for a universal "lowest" level – e.g. of elementary particles – sufficient for arbitrary future theories.

The usual arguments between protophysicists and "deutero-physicists" – if I may coin that term for the moment – using coordinate-systems with a spatial or even a spatio-temporal metric are beside the point as long as the problem of introducing a "metric" is exempt from a truly mutual discussion. What can be introduced rather than merely postulated on the basis of elementary common human experience (still ambiguous relative to the singular – general bifurcation) will lend itself to the discrimination of (empirical) actualisations from (rational) schemata for any preaction.

Universität des Saarlandes,
Saarbrücken

NOTES

[1] Cf. for the terms together with the idea B. Russell, 'Knowledge by Acquaintance and Knowledge by Description', *Proc. Arist. Soc. N.S.* **11** (1910/11); for further relating this distinction to the distinction of presence and absence of objects (expounded during the work in the research project on 'Wissenschaftssprache versus Umgangssprache. Probleme des Aufbaus einer Wissenschaftssprache in Literatur- und Kunstwissenschaft' conducted by D. Gerhardus and K. Lorenz, sponsored by the Deutsche Forschungsgemeinschaft from fall 1977 to spring 1980), the two steps of a primary and a secondary dialogue-situation in K. Lorenz, *Elemente der Sprachkritik. Eine Alternative zum Dogmatismus und Skeptizismus in der analytischen Philosophie*, Frankfurt 1970, become relevant.
[2] Parts of the following exposition derive from a further elaboration of parts of the paper by the author, 'The Concept of Science. Some Remarks on the Methodological Issue 'Construction' versus 'Description' in the Philosophy of Science', in *Transcendental Arguments and Science* (P. Bieri, R. P. Horstmann, and L. Krüger, eds.), Dordrecht 1979.
[3] Cf. H. Reichenbach, *Experience and Prediction*, Chicago/London 1938, esp. ch. I, and compare with the context of Leibniz's terms as expounded e.g. in H. Hermes, 'Die ars inveniendi und die ars iudicandi', *Studia Leibnitiana Suppl.* III, Wiesbaden 1969.
[4] A term used by W. James for his version of pragmatism, which is exactly in line with the claim just made, cf. the collection of essays in *The Philosophy of William James* (W. R. Corti, ed.), Hamburg 1976.
[5] Of course, the radical empiricism of W. James may be included, too, since this issue can be dealt with quite independently from the dispute between James and Peirce on the meaning of the term 'pragmatism'; cf. for support, e.g., Peirce's argumentation against first intuitions to secure cognition in 'Questions Concerning Certain Faculties Claimed for Man', C. S. Peirce, *Collected Papers* I–VI (ed. Ch. Hartshorne and P. Weiss), Cambridge, Mass. 1931–35, 5.213 ff.
[6] This derives from the fact that Kant never disputes the reality of knowledge, i.e. of Newtonian physics, but tries to clarify the conditions of its possibility; cf. the relevant exposition in the last chapter (§15) of J. Mittelstrass, *Neuzeit und Aufklärung. Studien zur Entstehung der neuzeitlichen Wissenschaft und Philosophie,* Berlin/New York 1970.
[7] H. Wohlrapp, 'Analytischer versus konstruktiver Wissenschaftsbegriff', revised version, in *Konstruktionen versus Positionen. Beiträge zur Diskussion um die Konstruktive Wissenschaftstheorie* (ed. with an Introduction by K. Lorenz), Berlin/New York 1979, Band II (Allgemeine Wissenschaftstheorie), 348–377.
[8] The latest account in my introduction to the reprint of R. Gätschenberger, *Zeichen,*

die Fundamente des Wissens, Stuttgart 1977; at the same place attempts to relate this approach with ideas of the symbolic interactionism as developed by G. H. Mead and of the genetic epistemology by J. Piaget; cf. also my essay 'Sprachphilosophie', in *Lexikon der germanistischen Linguistik* (ed. H. P. Althaus, H. Henne, and H. E. Wiegand), Tübingen [2] 1980, 1–28.

[9] Under the heading 'Property and Substance' being the terms for repeatable and non-repeatable entities respectively, the same issue is at stake when R. M. Rorty in his paper on 'The Subjectivist Principle and the Linguistic Turn' (in *Alfred North Whitehead. Essays on his Philosophy* (G.L. Kline, ed.), Englewood Cliffs, N.J. 1963) discusses – and refutes – the attempts of A. N. Whitehead to evade well-known epistemological dilemmas deriving from the singular-general dichotomy, if this dichotomy is correlated in a straightforward way, i.e. without using linguistic analysis, with the body-mind dualism.

[10] This idea is basic, already, to Peirce's treatment of the sign-process in the general framework of his theory of categories; cf. Lecture IV on 'The Reality of Thirdness' (*Collected Papers* 5.93–5.119) among the 'Lectures on Pragmatism' (the singular being existent only, is an object of knowledge through that which is real, i.e. the universal).

[11] Representative is the treatment in the last chapter of W. V. O. Quine, *Word and Object*, Cambridge, Mass. 1960, §56 (Semantic Ascent).

[12] Cf. the sophisticated treatment of the "empirical core" (= empirical content) of a theory via Sneed's criteria of theoreticity as expounded e.g. in W. Stegmüller, *Probleme und Resultate der Wissenschaftstheorie und analytischen Philosophie, Band II* (*Theorie und Erfahrung*), 2. Halbband (Theorienstrukturen und Theoriendynamik), Kap VIII, Heidelberg/New York 1973.

[13] Representative is the treatment of D. Davidson in his paper 'Actions, Reasons and Causes', in *The Journal of Philosophy* **60** (1963); the incorporation of both the mental and the physical aspect into the concept of action which is a natural consequence of the approach via preactions as proposed here makes it possible for Davidson to defend the much disputed Aristotelian claim of a causal connection between reason and action.

[14] Cf. P. F. Strawson, *Individuals*, London 1959.

[15] Cf. the paper by the author, 'On the Relation Between the Partition of a Whole into Parts and the Attribution of Properties to an Object', in *Studia Logica* **36** (1977).

[16] John Rawls, *A Theory of Justice*, London/Oxford 1973.

[17] Cf. K. E. Tranøy, 'Norms of Inquiry: Rationality, Consistency Requirements and Normative Conflict', in *Rationality in Science, Studies in the Foundations of Science and Ethics* (ed. R. Hilpinen), Dordrecht/Boston/London 1980, 191–202.

[18] Cf. one of the earliest versions in 'How to Make our Ideas Clear' (*Collected Papers* 5.388–5.410): "The opinion which is fated to be ultimately agreed to by all who investigate, is what we mean by the truth, and the object represented in this opinion is the real." (5.407).

[19] The pragmatic concept of truth is being developed in the game-theoretic approach of dialogic logic – cf. the collection of essays in P. Lorenzen and K. Lorenz, *Dialogische Logik*, Darmstadt 1978–; the semantic concept of truth uses the well-known model-theoretic approach initiated by A. Tarski in his classical paper 'The Concept of Truth in Formalized Languages', cf. A. Tarski, *Logic, Semantics, Metamathematics. Papers from 1923 to 1938*, Oxford 1956, 152–278.

[20] Cf. the exposition in the paper by Lars Bergström, 'Some Remarks Concerning Rationality in Science', in *Rationality in Science, Studies in the Foundations of Science and Ethics* (ed. R. Hilpinen), Dordrecht/Boston/London 1980, 1–11.

[21] This refers back to the Aristotelian distinction of actions as means for some outside end (making – ποίησις) and as ends in themselves (doing – πρᾶξις), discussed e.g. in *Eth. Nic.* A and Z4.

[22] Cf. the discussion of the interrelation between explanation and induction in C. G. Hempel, *Aspekte wissenschaftlicher Erklärung,* Berlin/New York 1977 (German translation of a revised version of the last chapter of C. G. Hempel, *Aspects of Scientific Explanation and Other Essays in the Philosophy of Science,* New York 1965).

[23] Cf. for comparison the related remarks on the difference between descriptive and explanatory adequacy of a theory, here: of linguistics, in N. Chomsky, *Aspects of the Theory of Syntax,* Cambridge, Mass. 1965, Chap. I (Methodological Preliminaries).

[24] On the alleged interdependence of a constructive and a normative approach to science cf. F. Kambartel and J. Mittelstrass, (eds.), *Zum normativen Fundament der Wissenschaft,* Frankfurt 1973, espec. the essays by J. Mittelstrass ('Das praktische Fundament der Wissenschaft und die Aufgabe der Philosophie'), P. Janich ('Eindeutigkeit, Konsistenz und methodische Ordnung: Normative versus deskriptive Wissenschaftstheorie zur Physik') and O. Schwemmer ('Grundlagen einer normativen Ethik'). The independence of 'is' and 'ought' is usually taken for granted, formalized as non-validity of $\Delta! A \prec A$ in deontic logic (cf. P. Lorenzen, *Normative Logic and Ethics,* Mannheim 1969, p. 70f); and attempts to question the is-ought gap get criticized even by other opponents of the analytic approach, cf. e.g. K.-O. Apel, who in his detailed discussion 'Sprechakttheorie und Begründung ethischer Normen' (in *Konstruktionen versus Positionen. Beiträge zur Diskussion um die Konstruktive Wissenschaftstheorie* (ed., with an Introduction by K. Lorenz), Berlin/New York 1979) takes pains to refute J. R. Searle's claims that there exist nontrivial logical relations among is- and ought-sentences. For further discussion cf. W. D. Hudson (ed.), *The Is-Ought Question,* London 1969.

[25] For further constructions in order to reach the usual level of syntactic differentiations, cf. K. Lorenz, 'Words and Sentences. A Pragmatic Approach to the Introduction of Syntactic Categories', in *Communication and Cognition* **9** (Gent 1976).

[26] The grades of theoreticity which serve as a kind of measure for the distance to common-sense experience (relative to some natural language system) are discussed by W. V. O. Quine, 'Grades of Theoreticity', in *Experience and Theory* (L. Foster and J. W. Swanson, eds), Cambridge/Mass. 1970.

CARL FRIEDRICH GETHMANN

THE PHILOSOPHY OF SCIENCE AND ITS LOGIC

In this article* I would like to investigate the question as to which logic the philosophy of science *should* employ as an instrument in the reconstruction of the sciences. In accordance with the tradition of modern logic following Frege and Russell the question also implies a more narrow understanding of "logic", according to which logic embraces the context-free and individually *in*variant rules of deduction. Methodical rules, for example, remain outside the domain of investigation. The expression "the logic of the philosophy of science" is therefore not to be regarded as being parallel to such terms as "the logic of scientific discovery" or the "logic of the social sciences."

The question presupposes, moreover, *that* the philosophy of science, however it may define its aim (as analysis, construction, reconstruction, reflection or otherwise) makes use of an organon of analysis etc., the reason being that *in scientific discourse* scientists rely on such a "rule-know-how" (in a manner which requires clarification). In what follows this presupposition will not be made the object of comprehensive attempts at justification although aspects of this basic supposition with regard to the aims of the philosophy of science, and some of its consequences, will be touched upon in passing.

The question can only be meaningful if there are logic*s* and, what is more, in a competitive sense of the plural. A non-competitive use of the plural would be one which merely designated the fact that there are various logical calculi; for example, propositional logic, predicate logic and modal logic, which differ from one another in their syntax and semantics but which cannot in the methodological sense be regarded as alternatives. One must admittedly make decisions regarding their application but the use of one form of calculus does not exclude the employment of another. It is rather the case that they may be incorporated in a methodological 'continuum' on the basis of their power of expression.

In contrast to this we refer to logic*s* in the sense of *competing calculi* which confront one another. The discussion between classical and intuitionistic logic which follows is initially intended to be an example

Robert E. Butts and James Robert Brown (eds.), Constructivism and Science, 19—45.

of this. In saying this we have already intimated that, in the final analysis, the relationship between classical and intuitionistic logic is not merely to be regarded as the relationship between two calculi (and certainly not as orthodox and heterodox variants) but between two *conceptions of logic* (formalism versus constructivism). The reason for taking this view is that immanent to both systems of logic is a completely different attitude to the problem of the foundations of logic. In the face of this difference the fact that, in the resultant classical calculi, more rule schemata are valid than in the intuitionistic calculi, is in itself without significance.

In the end the difference between classical and intuitionistic logic is, therefore, an *example* in the sense that it is not being suggested that it is merely a matter of this alternative. Instead, we intend to demonstrate that the discussion which has taken place regarding these logics is an adequate occasion for considering the conceptions of logic underlying them. Therefore we regard the question posed at the beginning of this article as a necessary one and hence as neither trivial nor already decided.

1. THE LOGICAL DOGMATISM OF RECENT ANALYTICAL PHILOSOPHY OF SCIENCE

In the context of the questions being discussed, logical dogmatism is to be understood as the view that the choice of a logic on the part of the philosophy of science requires no discussion, that is to say, it raises no problems of justification since it is only a formal-semantically conceived classical logic which can be considered as a suitable instrument for philosophical-scientific analysis.

A logical dogmatism of this sort is noticeable in recent analytical philosophy of science. By way of example we may take Stegmüller's comprehensive account of the 'Problems of and Results of the Philosophy of Science and Analytical Philosophy'. Since, according to Stegmüller, the philosophy of science has only *one* method, namely the application of logic,[1] he prefaces this work with a chapter entitled the 'ABC of Modern Logic and Semantics', which bears the stamp of logical dogmatism in both conception and execution.[2] An expansion of this chapter to take in further logic*s* is only recognised in the non-competing sense of the plural.[3]

The problematic character of this logical dogmatism as shown by numerous other analytical philosophers of science, especially in connection with the assumption of the 'in principle' correct intuitions of the specialist philosopher, still has to be discussed. At this point one should first make the *historical comment* that logical dogmatism by no means accords with the views of the founders and early proponents of analytic philosophy of science.

Although Brouwer's criticism of classical logic in the 1920s had certainly not yet led to a form of logic which was able to compete with the classical account in *Principia Mathematica*, many members of the Vienna Circle adopted a decidedly cautious but often sympathetic attitude to the question of the 'correct' logic.[4] Some members even endorsed Brouwer's epistemological objections under the rubric of '*Finitism*'. In their programmatic "The Scientific Conception of the World: The Vienna Circle" Carnap, Hahn and Neurath write with reference to the difference between logicism (Frege, Russell), formalism (Hilbert) and intuitionism (Brouwer):

The debates are followed with great interest in the Vienna Circle. Where the decision will lead in the end cannot yet be foreseen; in any case, it will also imply a decision about the structure of logic; hence the importance of this problem for the scientific world-conception.[5]

Even this reserved statement contains a criticism of the view, put forward particularly clearly by Wittgenstein in the *Tractatus,* that classical logic is to some extent isomorphic with the world and hence structurally immanent to language. Even H. Hahn, who maintained the validity of classical logic did not follow the naturalistic tendency of the *Tractatus* but characterized the rules of logic as conventional rules for the use of signs.[6] Hahn writes:

What logical deduction accomplishes, then, is this: it makes us aware of all that we have implicitly asserted – on the basis of conventions regarding the use of language – in asserting a system of propositions.[7]

O. Neurath and F. Kaufmann in particular, supported the finitistic and hence, implicitly, the intuitionistic view. Neurath emphatically includes attempts "to make mathematics finite, especially in applications to concrete events" among efforts to purge the sciences of theological-metaphysical residue.[8] In his book, *The Infinite in Mathematics*, Kaufmann had made it his principal task to criticize the assumption of the actual infinite and declares of such mathematical propositions as contain the notion of the indenumerably infinite: "Such propositions will turn out to be sham judgments and will have to be eradicated from mathematics."[9] Kaufmann makes it clear that Brouwer's criticism has decisively influenced his view even if he does not adhere to it in all of its details.[10]

Carnap, too, in his report on the genesis of his book, *The Logical Syntax of Language*, confirms that there was general sympathy within the Vienna Circle with finitistic ideas. Hence it initially had been his intention to develop only language I (which was constructed with finite means alone).[11] In order to assess Carnap's conception of the tolerance principle[12] in questions of logic (later to be called the principle of conventionality[13]) it should be noted that the work of A. Heyting had at the beginning of the thirties successfully produced a complete calculization of intuitionistic logic so that the question as to *which* system of logic the philosophy of science would use for purposes of analysis and reconstruction became inescapable. In the published version of *The Logical Syntax of Language* Carnap accordingly formulated two logical languages. The first was constructed to fulfill the essential requirements of the intuitionists by permitting only the application of definite linguistic methods; the second was based on classical logic. It is important that Carnap, although he did not share the reservations of the intuitionists about the permissibility of indefinite concepts,[14] did point out in his tolerance principle that the definition of the task of the logical construction of a scientific language does not *eo ipso* determine which logic is to be the logic of the philosophy of science. The introduction of the tolerance principle is a reaction to the question of the correct logic. "In logic there are no morals. Everyone is at liberty to build up his own logic, i.e. his own form of language, as he wishes."[15]

If one regards Carnap's conception as being symptomatic of the dominating tendencies within the Vienna Circle, this evidence shows *that the founders of the analytical theory of science did not regard themselves as logical dogmatists but as logical conventionalists* and, hence, it was left to the philosopher of science engaged in reconstruction to choose his own form of logic. In spite of the demand for tolerance, Stegmüller, for example, does not observe Carnap's tolerance principle in his account of the analytical philosophy of science.[16]

The logical dogmatism of recent analytical philosophy of science, however, may also be seen as a tacit admission that conventionalism with respect to logic, particularly for a discipline that employs logic as a tool, is not only didactically confusing, but also has a negative influence on the philosophy of science's task of clarification and enlightenment. The number of methodological disputes would increase rather than decrease by taking into account several systems of logic (the controversy between classical and intuitionistic logic is only, of course, intended to be one example). What is more, conventionalism entails not only tolerance towards other known forms of logic but must also take into account that a 'quite different' logic may soon be constructed which will then in the same way have to be permitted to take its place as "another voice in the concert." Carnap clearly reckons with such a possibility: "It could doubtlessly be the case that a system which deviates from the traditional form will prove useful for the foundation of the language of science."[17] If one is of the opinion – as many recent proponents of analytic philosophy of science assume in their work – that the competitive application of several forms of logic is inappropriate to the task of the philosophy of science, then the question of an appropriate logic is unavoidable. This question cannot, however, be decided by *logical* means alone. To this extent Carnap's tolerance principle is acceptable from the point of view of the logician, whose interests are not primarily instrumental. The philosophy of science, however, employs logic as an instrument for the solution of what must be considered a practical task not motivated by logic. Formulated as neutrally as possible, it is the task of understanding the sciences better with the aim of improving them.

Conventionalism in the sense of the tolerance principle is, therefore, unacceptable in those cases where logic is employed for the purpose of the reconstruction and analysis of scientific languages. The question as to whether the means are suited to the ends must be raised and answered. All the more so when philosophers of science who believe that logic is completely unsuitable for the complete or even partial reconstruction of the rationality structures of scientific knowledge enter the arena.[18] While logical dogmatism is uncritical, as it makes *de facto* decisions between possible alternatives without giving reasons, logical conventionalism is obviously unsuitable for the philosophy of science since it can lead to scientific controversies which have their origins not in the sciences themselves but in the instrument employed for their reconstruction. Hence the question formulated at the beginning of this article is decidedly relevant. The use of logic as an instrument of reconstruction in the philosophy of science does not suspend the question of justification, rather it makes it more urgent.

2. LOGICAL PROBLEMS IN ANALYTICAL PHILOSOPHY OF SCIENCE

Before this question of justification is taken further we should deal with the possible objection that an answer is without consequence for the work of the philosopher of science, since the choice of a type of logic has no influence on the problems with which he is dealing. On the basis of a selection of central problems (not only) of the philosophy of science, it will first be shown that the choice of a particular kind of logic has consequences for the formulation and solution of problems in the philosophy of science. That an important argument might be gained from this is clear from the following comment of Stegmüller's on intuitionistic criticism of classical logic:

> Should justified suspicion exist with respect to certain types of logical operations because these can be shown to be responsible for the appearance of antinomies, then we have to dispense with these operations, especially within the framework of a semantic metatheory.[19]

It may be supposed that Stegmüller would recognise such consequences not only for antinomies, that is, contradictions within a calculus but also for paradoxes, that is, contradictions between a calculus and knowledge otherwise based. This may be considered sufficient reason to limit the following examples to known *paradoxes in the philosophy of science* and hence to supply 'evidence' for the 'suspicion' that not only do they have something to do with the objects being examined but also with the instrument of analysis, the kind of logic used.

a. Within the framework of Hempel's theory of confirmation we have the *Raven paradox* – according to the clear account by W. Lenzen[20] – if one proposes the following conditions of adequacy.

I. For a hypothesis $\bigwedge x\,(Fx \rightarrow Gx)$ formulated in the form of a generalized subjunction an observation proposition of the form $Fa \wedge Ga$ is a confirmation. (For the hypothesis 'All ravens are black' the sentence 'This raven is black' is a confirmation.)

II. Logically equivalent hypotheses are confirmed by identical observational propositions and logically equivalent observational propositions confirm the same hypotheses.

III. The observation of objects not possessing the qualities named in the antecedent of the hypothesis has no confirmatory power with respect to the hypothesis.

The logical analysis now shows, in accordance with Lenzen's account, which follows Hempel, that (III) cannot be satisfied if (I) and (II) are valid. The adequacy conditions (I) to (III), which many scientists would surely feel to be an explication of their intuitive notion of scientific confirmation lead to the paradoxical result that, for example, the observational proposition 'This elephant is pink' is a confirmation for the hypothesis 'All ravens are black.' Lenzen demonstrates this by the following derivation:
According to (I) the following relation is a confirmation:

(1) $\quad \text{conf}\,(\neg Ga \wedge \neg Fa, \bigwedge x\,(\neg Gx \rightarrow \neg Fx))$

Logically the following equivalence is valid:

(2) $\quad \bigwedge x\,(\neg Gx \rightarrow \neg Fx) >< \bigwedge x\,(Fx \rightarrow Gx)$

Hence according to (II) the following relation is also valid as confirmation:

(3) $\quad \text{conf}\,(\neg Fa \wedge \neg Ga, \bigwedge x\,(Fx \rightarrow Gx))$,

which contradicts (III) and hence leads to the paradox. One can immediately see that this paradox only arises if we use classical logic. The equivalence (2) is not intuitionistically valid. It is rather the case that intuitionistically, the following relationships are valid:

(2′) $\quad \bigwedge x\,(Fx \rightarrow Gx) < \bigwedge x\,(\neg Gx \rightarrow \neg Fx)$

and

(2″) $\quad \bigwedge x\,(\neg Gx \rightarrow \neg Fx) >< \bigwedge x\,(Fx \rightarrow \neg\neg Gx)$

so that instead of (3) we would have the relation:

(3′) $\quad R\,(\neg Fa \wedge \neg Ga, \bigwedge x\,(Fx \rightarrow \neg\neg Gx))$

This is, according to (I), not a confirmation. $\neg\neg Gx$ is not, in every case, equivalent with Gx because – and the raven paradox is a good example of this – having reasons for not being able to deny the presence of a property G is not the same thing as having reasons for being able to assert G.[21]

Consider the variant of the raven paradox which arises via the equivalence:

(4) $\quad \bigwedge x\,(Fx \rightarrow Gx) >< \bigwedge x\,(Fx \vee \neg Fx \vee Gx \vee \neg Gx \rightarrow \neg Fx \vee Gx)$

which entails that any object not possessing the property F or possessing the property G satisfies the hypothesis:

(5) $\text{conf}\,(\neg Fa \vee Ga,\ \wedge x\,(Fx \rightarrow Gx))$

This variant is of logical interest since (4) is classically and intuitionistically valid, but not from the point of view of minimal logic. This variant gives grounds for us to wonder whether even intuitionistic logic is not still too strong for the reconstruction of problems of this sort.

These comments show that the particular variants of the raven paradox depend not only on the conditions of adequacy but also on the logical calculus used. In Lenzen's detailed discussion of attempts at a solution, however, no significance is attached to the question of the type of logic used. Moreover, in the preface to his investigations, he states (quite in the spirit of logical dogmatism):

At some points in this paper the formalization of certain statements is unavoidable for the purpose of precision. To this end we use a first order predicate logic as described, for example, in Kutschera/Breitkopf (1970).[22]

b. A second example of the dependence of problems of the philosophy of science on logic is to be found in the *paradox of dispositional predicators*. A paradox (first noticed by Carnap[23]) in the definition of dispositional predicators, in the use of which, it may be assumed, most scientists see no intuitive problems, arises under the following conditions: If Dx is the predicate "x is soluble in water", where Wxt stands for "x is put into water at time t" and Lxt for "x dissolves at time t", the following definition suggests itself:

(6) $Dx \leftrightharpoons \wedge t\,(Wxt \rightarrow Lxt)$

If one now assumes of any object a, that it is *not* put into water at time t then we must *logically* derive the curious conclusion that it is soluble:

(7) $\wedge t\,(\neg Wat) \prec \wedge t\,(\neg Wat \vee Lat)$

(8) $\wedge t\,(\neg Wat \vee Lat) \prec \wedge t\,(Wat \rightarrow Lat)$

(9) $\wedge t\,(Wat \rightarrow Lat) \prec Da$ according to (6)

As is well known very different strategies were adopted in order to try to solve this problem. Carnap first tried to partially characterize dispositional predicators by means of reductional propositions, which led to methodological problems. Later the dispositional predicators decisively motivated his decision to go over to a two-level scientific language. Other authors have tried to replace the subjunction in (6) or in other formulations by other linguistic relations such as causal implication or subjunctive conditional sentences, the reconstruction of which led to problems of their own. It was probably disregarded that the paradox can only be generated classically and intuitionistically and not by using minimal logic. Using minimal logic, not (8) is valid but for instance

(8′) $\bigwedge t\,(\neg Wat \vee Lat) < \bigwedge t\,(Wat \rightarrow \neg\neg Lat)$

Again one would first have to discuss which logic is the adequate logic for such problems. In Stegmüller's account of the problems, an examination of the type of logic used is, however, not proposed as a subject for discussion.[24]

c. *The paradox of derived obligation* following Prior[25] serves as a third example within the framework of deontic standard logic, whose significance lies, for example, in the reconstruction of legal argumentation. According to deontic logic, without contradicting the intuition of the specialist, if it is not permitted to do A, it is also not permitted both to do A and not to do B

(10) $\neg PA < \neg P\,(A \wedge \neg B)$

If we now replace permission by obligation in accordance with:

(11) $PA \leftrightarrows \neg O\,\neg A$

and if we further allow, for the external negation, double negation elimination:

(12) $\neg\neg\Omega\,A < \Omega$ (where Ω stands for any deontic operator)

and further apply the definition of the prohibition in accordance with:

(13) $FA \leftrightarrows O\,\neg\,A_i$

and then finally apply as valid in propositional logic:

(14) $\neg(A \wedge \neg\,B) >< A \rightarrow B_i$

we then have the paradoxical expression derivable from (10):

(15) $FA < O\,(A \rightarrow B)$.

This states that, in the case of someone's not being allowed to do A, there is an obligation to do any B, if A were done. For example, if it is forbidden to murder, then one is obliged, for example, to rob the murdered man, if he has been murdered (and this even applies to the murderer.)

As is well known, (14) is not valid in intuitionistic logic. This is true; (15) is intuitionistically valid independently of (10) if the deontic modus ponens is applied:

(16) $OA, A \rightarrow B < OB.$

The required theorem

(17) $\neg A < A \rightarrow B,$

on the other hand, is intuitionistically valid but not in minimal logic. This variant certainly does not yield the same 'paradox effect' as the derivation based on (10). The former produces a paradoxical expression out of a nonparadoxical one, whereas (15) is an expression of the paradoxical character that may be denied since $A \rightarrow B$ is not intuitionistically equivalent to $\neg (A \wedge \neg B)$.

It should be noted that the protological reconstruction of the conditioned obligation by means of $O(A \rightarrow B)$ is not held to be adequate by many logicians, for example, Prior[26] and Lorenzen.[27] From the alternative $A \rightarrow OB$ we may obtain classically and intuitionistically, but not with minimal logic:

(18) $\neg A < A \rightarrow OB.$

It still remains to be discussed whether one should not declare (12) invalid and hence be an intuitionist not merely in respect of the embedded predicate calculus. This would mean that one could distinguish between two reconstructions of the prohibition, namely:

(19) $F_1 A \leftrightharpoons O\neg A$

and

(20) $F_2 A \leftrightharpoons \neg\neg O\neg A \leftrightharpoons \neg PA.$

Taking (20) as an 'intuitionistic prohibition' one could prove (10) to be identical in meaning only with

(21) $F_2 A < F_2 (A \wedge \neg B).$

These examples are, of course, by no means adequate to substantiate the assertion that the philosophy of science should make use of intuitionistic or minimal logic instead of classical logic. There are, however, variants of paradoxes which arise essentially through the use of particular logics and (given that the other methodological criteria such as the conditions of adequacy remain constant) would not arise if another form of logic were used.[28]

It will suffice to draw from these examples the conclusion *that classical logic is not so successful as a reconstruction instrument for the philosophy of science that the question of the justification of the application of a logic could (pragmatically) be said to be superfluous*. It is rather the case that, when paradoxes arise, especially such as appear very curious to scientists' intuitions (regarded by analytical philosophers of science as being quite reliable), not only the question of the conditions of adequacy but *also* the question of the choice of the 'correct' logic must be raised.

3. ON THE PROBLEM OF THE JUSTIFICATION OF A 'LOGIC OF THE PHILOSOPHY OF SCIENCE'

If the choice of a logic for dealing with problems of the philosophy of science has far-reaching consequences, then the principle of tolerance cannot be understood in the sense that this choice may be left to the discretion of the individual scientist. Whereas the logician cannot be criticised for the choice of the type of logic with which he wishes to concern himself, the applied logician must consider the reason why he is choosing one logic rather than another. It is clear that these reasons cannot lie in logic itself. It is not immediately obvious whether a logic is a suitable instrument or not. Criteria like "convenience" and "beauty" (Quine[29]) are therefore of no significance. The question as to which logic one chooses is, as Carnap formulates it, merely a question of the "practical reasons for preferring one or the other form for given purposes".[30] It is worth noting (for an understanding of the historical development of the tolerance principle) that Carnap understands his formulations as a consequence of the tolerance principle.

F. Waismann made a very similar comment in his essay "Are There Alternative Logics?"

> What we must understand is that the choice between distinct systems of logic is not decided by logic. It may happen that one system is suitable for this, the other for that purpose.[31]

Hence the question formulated at the beginning of this paper is, according to the dominant tendency among the authors of the Vienna Circle, not one that – it should now be said terminologically – can be settled by the substantiation of assertions, but by the *justification* of directions. Such a justification takes the form of demonstrating the choice of a logic to be a (perhaps better) means for the attainment of an end, namely the analysis and reconstruction of sciences.

These comments show that conventionalism does not, strictly speaking, entail the suspension of any question as to the foundation of a claim to validity with respect to the application of logic, but only that the question, in the sense of being the substantiation of an assertion, is declared to be senseless. It also becomes clear at this point that the use of the term '*convention*' is ambiguous in this context. 'Convention' may mean any agreement not accessible to any demands for substantiation or justification. It may, however, also be understood to mean the rules presupposed in any act of communication.

The quotation from Hahn – that logic represents, on the basis of conventions in the use of language, that which is implicitly concomitantly asserted[32] – points in the direction of this latter view. For this state of affairs one should probably rather use the term '*presupposition*' instead of convention.

After these considerations and terminological elucidations, it may be said, following the view of the members of the Vienna Circle, that the choice of a logic on the part of the philosophy of science is accessible to attempts at justification, which must be based on presuppositions of the use of language (here: by scientists). In short, the choice of the correct logic is not a descriptive but a prescriptive (regulative) problem. The clearest statements in this direction are by F. Waismann in the article already mentioned. After a precise differentiation and discussion of various types of logic he maintains (concerning a logic with *tertium non datur*) that such a logic gives expression to our striving after decidability. This striving is itself an attitude "imposed on us by the conditions of our life." Nevertheless we cannot overlook the fact that we occasionally also have an interest in leaving things undecided. In a clear reference to ideas of Wittgenstein in the *Philosophical Investigations* Waismann then noted that:

> This seems to point out a relationship between our way of thinking and a pattern of life. It may be that there are different types of behaviour (such as resoluteness, hesitance)

which make for different types of logic. The structure of our logic reflects in some way characteristic attitudes of our life. All this would open up a vista of something like a social substructure of logic.[33]

These remarks of Waismann's adumbrate a program for a *pragmatic justification of logic* which takes as its starting point the presuppositions of speech acts and actions in the world in which we live.

In order to put such a concept into practice, a simple procedure would seem to suggest itself; if the choice of a logic is to be justified by its purpose, then it seems that the philosopher of science would be advised to choose a logic that is employed at least *de facto* by scientists. This approach seems especially apt if one agrees with Stegmüller that the intuitions of specialist scientists are, in principle, correct.[34] If one examines the question more closely, however, a procedure of this sort turns out to be unworkable. Which logic scientists are using is not normally evident. It is precisely for that reason that analysis and reconstruction are necessary in the philosophy of science. Both scientists and non-scientists normally have at their disposal an argumentation-know-how, which they have acquired in the process of socialization, without being able to demonstrate explicit knowledge of the rules (know-that). Hence, in order to reconstruct the rules 'employed by' the specialist scientist, one already needs a logic, and the problem of justification again arises. As a rule even explicit questioning cannot help us here because the questions of the logician would be incomprehensible to the scientist. What scientist would be able to say whether he was arguing with or without *ex falso quodlibet* (that is, what notion of negation he was employing) or whether in using quantors he was using the variables substitutionally or referentially?

Even mathematics cannot help us here. It might be hoped that a mathematician has an explicit system of rules with regard to the methods of proof which he is using. Both classical and intuitionistic logicians, can, however, be found among mathematicians. What does it mean here, to use the same logic as the specialist scientists employ? A philosophy of science which leaves everything as it is, that is to say, only reports to mathematicians on its differences of opinion, would obviously be superfluous.

These remarks should make it clear why it may be regarded as hopeless to regard the problem of the justification of logic as a *descriptive one*. The choice of a particular logic cannot be justified by its being represented as the appropriate description-analytical instrument of a science, for, in order to describe the rationality structures of a science, one already requires a logic. This remark is, moreover, not only directed at philosophers of science who display a logical dogmatism to classical logic. The factual use of the *tertium non datur* by scientists cannot be regarded as an argument in favour of classical logic as the logic of the philosophy of science, nor can the factual existence of problems such as a proof of existence for an odd whole number be seen as an argument against classical logic. The critical treatment of such examples by the philosopher of science depends much more on the type of logic that he has already found most expedient for the reconstruction of the science in question.

The Logic of the Philosophy of Science is not Derivable from the Sciences Independently of a Logical Apparatus. Criticism of the view of logic as a descriptive theory of structure can also be formulated by considering *problems of definition* in the context of the definition of scientific rationality. 'Logicity' is a prior requirement as a means of definition for 'scientific rationality' so that 'logic of science' cannot conversely be defined by recourse to a notion of scientificness. In general philosophy of science, to suggest a notion of scientific rationality (for the purpose of distinguishing between science and pseudo-science, for example) one must have at one's disposal general rationality structures among which the rules of rational argumentation may surely be numbered. The isolation of logical rules from the factual practice of argumentation in the world in which we live (which follows logical and other rules) can therefore, for methodological reasons, not have recourse to a clarified notion of scientificity.

The considerations outlined so far all lead to the conclusion that the philosophy of science would be advised to understand logic not as *descriptive theory of structure* but as a *regulative theory of action* (that is, of scientific speech acts). For this purpose logic must already be preceded by elements of an understanding of linguistic acts relevant to argumentation, *a pragmatics of argumentation in the life-world.* With the help of the latter we then have to reconstruct initial logical rules which are not bound to a calculus. This part of a more comprehensive theory of argumentational speech can, therefore, – in accordance with a terminological suggestion by P. Lorenzen – be called "protologic".[35]

4. ON THE CONSTRUCTION OF A PRAGMATICALLY ORIENTATED PROTOLOGIC

It is assumed by most logicians and philosophers of science that modern standard logic is, in a manner which still has to be clarified, a theory of argumentation, especially of scientific argumentation. This assumption is, however, by no means obvious. Argumentation as a special form of linguistic action incontrovertibly embraces such types of action as asserting, doubting, and assenting. The standard accounts of logic, however, generally describe logic as a theory of the structural relations between propositions (distinctions such as classic-intuitionistic being unimportant here). Accordingly, in logic one can clarify the question as to whether there is an inferential relationship between "Fido is a brown dog" and "Fido is brown" but not whether there is one between "A assents to p" and "A asserts that p". Stegmüller too, for example, assigns logic the task of formulating the rules of correct argumentation. He declares, on the other hand, that these rules only apply to 'artificial languages' and not to the speech acts of a natural language. The task of translating between both languages is assigned to a separate discipline – rhetoric, whose functions have unfortunately been forgotten until the recent past.[36]

Hence, the problematic of translation inherent in this standard logic approach, which is closely connected with the understanding of logic as a descriptive theory of structure, fails to provide a solution to the problem of justification, which cannot be avoided by the pragmatically orientated philosopher of science: namely, why this and not another or indeed any artificial language (better, calculus) should be able to reproduce the rules of scientific *argumentation*, that is, actual speech acts (at least in part). It is for this reason that the acknowledgement of the problems of justification in conjunction with the notion of logic as a regulative theory of action compels us to give up the formal semantic and propositional (*satzlogische*) view of logic and instead, attempt a pragmatic construction of logic which has as its starting point speech acts in their argumentative context. Accordingly logical rules are special rules of argumentative language.

At this point it becomes clear why the dispute between classical and intuitionistic logic can only serve as a provisional example in order to demonstrate that the problem of justification is at all necessary. If the problems of justification are ultimately seen in so acute a light, the more fundamental alternative of logic as the science of the laws of propositional structures, or logic as the pragmatics of speech acts, acquires a methodologically crucial significance. In the pragmatic view a conception of logic in the tradition of the *logic of rules*[37] is far more important than the question of whether one holds classical logic or an intuitionistic variant of this logic to be primary, although a consistent methodical construction of the procedure of justification should also supply criteria for the settling of such questions.

As far as the philosophy of science is concerned, it should be especially emphasised that even in the sciences it is not the case that only *logical* rules are employed. Scientists also employ rules that are traditionally designated *rhetorical*. The validity of the latter was considered to be context-dependent and individually-variant. In contrast the *logical* rules could be understood as being valid independently of context and individual (in short, situationally invariant). This characterization also permits the further distinction of an intermediate form, contextually variant but individually invariant rules, which are only valid in certain areas of knowledge but are here universally valid. They may be called *topical* rules.[38] The deficiencies of the theoretical treatment of such rules can only be hinted at here. At this point it will suffice to emphasise situational invariance as the principal methodological criterion for the extrication of the *logical* rules. *The logical rules of scientific argumentation are precisely the situationally invariant rules of argumentative speech.* For the step-by-step methodical reconstruction of logical rules in this sense *three* phases can be distinguished. These are methodologically constructed on the basis of one another and are characterized by different methodological criteria of selection (principles).[39]

a. Acts such as assertions, promises, assents, entreaty, etc., are elements of language activity. Being able to execute such speech acts again and again implies having the relevant speech-act-schemata at one's disposal. Important elements of such schemata are performation and propositions of the latter, nomination and predication. The numerous speech-act-schemata which human beings have at their disposal can be methodically ordered according to the question as to which performations one must already have at one's disposal in order to learn others. Accordingly the constative type of speech act, for example, can be defined as that which is methodologically based on assertions. The articulation of speech-act schemata, moreover, represents in itself a possibility of normal language with whose aid the *comprehensibility* and *reliability* of speech is assured. These criteria are also of primary importance when typical successions of speech acts between speakers are to be characterized. For constative speech acts we can therefore formulate a standard form of substantificatory discourse in such a way that whereas it is not factually complied with, actual speech sequences can be reconstructed as being executed in accordance with such schemata. This first phase in the step-by-step introduction of individual speech act schemata and sequences of such will be called the "*schematization of discourse.*"

b. The formation of opinion in discourse constrained by the limitations of the time and mental energy available leads, from the point of view of the criteria of *abbreviation* and *simplification*, to the task of modifying discourse sequences while upholding their argumentative function. In order to effect this modification language practice once again supplies particular linguistic means whose use can be normed on the basis of their instrumental function (operator). Of fundamental importance for the understanding of the meaning of the operators are elementary discourse abilities originating in the life-world. In particular the following transitions of meaning from pragmatically practiced abilities to explicitly normed operators can be established:

practiced in the life-world: . . . of actions		methodically introduced: . . . of propositions
succession	$\Rightarrow$	subjunction
repulsion		negation
addition		conjunction, generalization
alternative		adjunction, particularization

The rules of meaning are pragmatically completely determined when their function in the abbreviation and simplification of discourse is known; when, thereby, the rules for the introduction and elimination of operators are known. The latter are, of course, in a methodological sense 'inverse' so that there is a possibility of reducing them by logical means.[40] Abbreviation and simplification rules, in so far as their situational invariance, i.e. their independence of individuals and context of discourse can be demonstrated, are *initial logical rules* known and employed before the formulation of an explicit calculus. The relevant phase of their methodical reconstruction can therefore be termed the "*logicization of discourse.*"

c. In accordance with the criteria of *completeness* and *reliability* (freedom from contradiction) the combinability and incompleteness of the set of logical rules discussed so far now gives rise to the methodological necessity of gaining an overview, which permits us to answer, for each of the rule candidates, the question of whether it is "logical" or not. The introduction of propositional and predicate variables[41] makes possible schematic calculation with rules so that we can decide in the case of any rule whether it is reducible to the initial introduction and elimination rules in a finite number of steps. The question of whether such a calculus is to be presented in the form of a calculus of natural deduction – a calculus of sequents in semantic tableaus or dialogue form – is, as a technical problem, a matter of secondary importance.[42] This section of the pragmatic justification of logic we shall call the "*calculization of discourse*". The calculization of discourse requires not only the demonstration that the linguistic instruments of the calculus permit us to express all pragmatically meaningful situations but also that in the calculus one and the same rule can never be permissible and impermissible at the same time.

As a result of this construction we have a methodical sequence[43] between minimal calculus, intuitionistic, and classical calculus in the following sense:

The six pairs of rules of the introduction and elimination rules methodically generate first the *minimal-logic-calculus*[44] since there is no protological justification for the *ex falso quodlibet* (EFQ). That is to say, the notion of the logically false can only first be introduced with reference to an existing calculus. If one already has the minimal calculus, the EFQ can be added as a further rule and one then has an *intuitionistic calculus.* Of this it may be said that it is *the* logical calculus if, with regard to the utterances to which it refers, no contextual commitments are entered into. If argumentation takes place in contexts in which each proposition to appear can either be asserted, giving reasons, or denied (that is to say that a 'truth-value' can be related to each proposition) and hence the *tertium non datur* can be thus added to the intuitionistic calculus, one then has a *classical calculus.* In accordance with the terminological suggestions made above, rule systems whose validity is dependent on their context are *topical*. The classical calculus is accordingly an initial topical arrangement achieved through methodical construction. Its contextual conditions are, moreover, satisfied in many areas of science so that no reservations about the use of classical logic can exist in them. By formal-pragmatic variation of the contextual assumptions further topical arrangements could be developed. We can merely hint at a program for a *formal topics* within which, by the use of suitable contextual assumptions, rules for argumentation referring to specific areas could be methodically developed around the core of formal (intuitionistic) logic. The example of the methodological relationship between minimal logical, intuitionistic and classical calculus already makes it possible to demonstrate that the application of a calculus can take place for reasons that arise from the methodological justification. This shows *how* a justification of the choice of a calculus as an instrument for the reconstruction of a scientific language can result (insofar as one has precisely such a calculus at one's disposal or the philosopher of science orders one from the logician).

The methodological conception developed here is structurally the opposite of the usual sequence of *syntax, semantics and pragmatics*. This 'formalistic' conception presents two decisive difficulties. On the one hand, conventional formal semantics is unsuitable for application in the context of justification as it already makes substantial use of the operators to be justified (the problem of circularity).[45] On the other, the discipline of pragmatics is always mentioned as a constituent of semiotic language construction but a discipline of this name does not exist within the semiotic framework. Hence it has become common to speak of pragmatics when formal semantics is expanded with indexical expressions and expressions for performative phrases.[46] In contrast to this, the construction of protologic (in three phases) and logic suggested here consistently assumes that logic is concerned with the stylized rules of argumentation.

"Pragmatics" is therefore methodologically prior to axiomatics. In axiomatics, however, signs which are in need of interpretation no longer appear, so that there is no longer any need for an independent discipline by the name of semantics.[47] Hence the conception of justification which has been outlined above avoids not only dogmatism with regard to classical logic but with regard to any logic. It is conventionalistic in the sense that it investigates those presuppositions of argumentation which scientists (and human beings in general) performatively recognise when they participate in argumentation. This conception implies that in a "possible" world in which there are beings who do not participate in argumentation (for example, because they are omniscient) or in which there are beings whose argumentation is not subject to the pressure of time and hence are able to discuss each elementary proposition asserted (for example, because they are immortal) the justification of a logic cannot succeed as it is not an instrument for a generally acknowledged end. It can be seen that logic (like topic and rhetoric) is something specifically human.

5. CONSEQUENCES FOR THE STATUS OF THE PHILOSOPHY OF SCIENCE

Even if methodological rules are conceded an independent status *vis-à-vis* logical rules and one, therefore, does not wish to go as far as Stegmüller in merely regarding the philosophy of science as applied logic,[48] one must conclude that the status of the philosophy of science is largely dependent on the understanding of one of its most important instruments – logic. If one is not prepared to be content with logical dogmatism and wishes to find a justification for the application of a logic, and if one further bears in mind that this justification can neither result from the introspection of the philosopher of science nor by research into the opinions or behaviour of specialist scientists, then the analytical-descriptive position must appear questionable not only for logic but for the philosophy of science which employs it. In principle, the correct intuitions of specialist scientists (to return to the formulation of Stegmüller already criticized above) have no operable meaning if the question of what the intuitions of the specialists are results from the reconstructive-critical work of the logician or philosopher of science. This work, moreover, already needs a logic as a working instrument. The "fact science"[49] is not a possible methodological starting point for the philosophy of science (even if it is often a heuristic one), since what science can mean first results from critical reconstruction. For this purpose at least (as *one* characteristic of scientificness) we already require a clarified notion of logic. If one therefore accepts that there are compelling reasons in favour of overcoming logical dogmatism through a conception of logic as a regulative theory of action, the descriptivism of the philosophy of science can no longer be adhered to.[50]

One consequence of the reconstructive-critical view of the theory of science is that Stegmüller's demarcation between the *philosophy of science* and *theory of knowledge* can no longer be upheld. Stegmüller sees the difference as being that the philosopher of science assumes that a rational reconstruction of the sciences is possible while the theory of knowledge questions even these assumptions.[51] A 'rational' reconstruction can, however, only be a reconstruction with the aid of a logic which – as has been demonstrated – is in need of justification. Hence, if the philosopher of science wishes to avoid logical dogmatism, the possibility of rational justification is just as open to debate. The question of validity ('*quid iuris*') is the central question of any philosophy of science which considers logical dogmatism, like any dogmatism, to be incompatible with that claim to critique to which it owes its own existence.[52]

The conception of a pragmatics of scientific argumentation which I have outlined here must also lead to a revision of the relationship between the *philosophy of science* and *the sociology of science* (the science of science). The traditional division of labour between (analytical) philosophy of science and (functionalistic) sociology of science is based on the dichotomy between cognitive ('internal') and social ('external') structures of the sciences. A pragmatics of scientific argumentation, in the sense in which we have been using it also permits, however, the critical reconstruction of social structures, for example, scientific institutions. Within the framework of a pragmatics of scientific argumentation we therefore have at our disposal a basic terminology and a set of rules which may be employed both for the formulation of cognitive structures, for example, forms of scientific *systematization* (theories, explanations, prognoses etc.) and also of social structures, for example, forms of scientific institutionalization. Scientific discourse as a pragmatic basis for scientific systematization and institutionalization is just as 'cognitive' as it is 'social'. In such discourse individuals largely interact by virtue of the fact that claims to validity are raised, doubted, and redeemed or rejected.[53] The institutional approach arises from the consideration that, in scientific discourse, it is indeed the case that, in principle, every claim to validity can be doubted and its verification attempted but not all claims at once, since discourse is rendered possible by regulative and constative prediscursive agreement. Forms of institutionalization arise when recognition of such prediscursive agreement is conceded for a relative period and defended with sanctions, primarily with respect to qualification and reputation; the self-definition of scientific communities is thus made possible.[54] The genesis of scientific institutions can therefore be reconstructed as the pragmatic conversion of assertions or rules requiring justification into *de facto* indubitable propositions. The *reconstruction* of such forms of institutionalization can only take place with the aid of those pragmatic instruments which also permit the reconstruction of forms of systematization. The *criticism* of scientific institutions can, however, no longer be conducted with these tools. At this point we would have to deal with the 'the philosophy of science and its ethics'.

APPENDIX (1984)

Induced by Section 2 of the essay above, R. Hegselmann and W. Raub scrutinized the extent to which paradoxes of the philosophy of science depend on a particular logical calculus.[55] For the examples of the paradox of dispositional predicators, the paradox of Goodman, the paradox of obligation and the Raven paradox, the authors demonstrate that the usually discussed variants of paradoxes only arise if derived on the basis of classical logic. In this respect they confirm the thesis maintained above:

> There are, however, variants of paradoxes which arise essentially through the use of particular logics and (given that the other methodological criteria such as the conditions of adequacy remain constant), would not arise if another form of logic were used.[56]

Hence R. Hegselmann and W.Raub introduce the distinction between a strong and a weak dependence on logic in order to point out that for all paradoxical expressions derived by classical logic one can find a variant arising on the basis of intuitionistic or minimal calculus. The corresponding transformation can be attained by adding the double negative elimination as antecedent (for the case of transforming classical into intuitionistic variants) or by replacing predicators of the form Gx by $\neg G'x$ (for transforming intuitionistic variants into minimal ones). Thus far the results of the examination are logically trivial. But if one would conclude from this 'translatability' that the choice of the logic did not matter, one would presuppose that those assumptions were valid. Only if, for example, the law of double negation elimination is valid are the correspondent expressions equivalent and do we have in a certain sense the 'same' paradox. But to claim these assumptions means to be a partisan of classical or intuitionistic logic respectively. Only for the classic is, for example, the definition

$$A \leftrightarrows \neg\neg A$$

non-creative, and only for the intuitionist, for example, the definition

$$B \leftrightarrows B \vee \curlywedge.$$

To discuss the creativity of those definitions means to discuss the questions of semantic theory which have to be clarified before one can choose in a justified manner the classical, intuitionistic or minimal calculus (or others) as the logic of the philosophy of science.[57] Hence, the uncritical maintenance of classical logic remains 'logical dogmatism' for which R. Hegselmann and W. Raub set a further example.

NOTES

* This Paper was first presented at the DFG-colloquium "Tasks and goals of the philosophy of physics", 22/23 September 1984 at the University of Konstanz.

1 *Vol. IV/1*, 7–8.

2 *Vol. I*, 1–71.

3 *Vol. IV/1*, 10; *Hauptströmungen. Vol. II*, 147–220.

4 In particular they referred to "De onbetrouwbaarheid", "Intuitionism", "Intuitionistische Mengenlehre", "Über die Bedeutung".

5 "Wissenschaftliche Weltauffassung: Der Wiener Kreis (The scientific Conception of the world: The Vienna Circle)", 311.

6 "Logic", 152–158.

7 *Op. cit.*, 157.

8 "Ways of the Scientific World-Conception", 43.

9 *Op. cit.*, 13, note 2. cf. pp. 3.

10 *Loc. cit.* 52–58.

11 "Intellectual Autobiography", 55–56.

12 *The Logical Syntax of Language*, 51.

13 *Introduction to Semantics*, 247.

14 *The Logical Syntax of Language*, Sec. 43–45.

15 *Loc. cit.* 52.

16 By addressing the reproach of intolerance to the modern 'finitists', that is, the constructivists of the Erlangen School (*Vol. IV/1*, 25–26).

17 *Foundations of Logic*, 129.

18 E.g. see S. Toulmin's distinction between logicality and rationality of science (cf. "From Logical Systems", *Human Understanding*).

19 *Das Wahrheitsproblem*, 250 (my translation).

20 *Theorien der Bestätigung*, 128 ss.

[21] See C.F. Gethmann, *Protologik*, 138–141.
[22] *Theorien der Bestätigung*, 11 (my translation). For a general methodological critique of the analytical theory of confirmation see H. Tetens, *Bewegungsformen*, 144–159.
[23] "Testability".
[24] *Probleme und Resultate. Vol. II*, 213–238.
[25] "The Paradoxes".
[26] *Formal Logic*, 224–225.
[27] *Konstruktive Logik*, 122–123.
[28] A misleading reference to Goodman's paradox in the German text has been deleted thanks to the examination of R. Hegselmann and W. Raub (see Appendix).
[29] *Philosophy of Logic*, 87, 88.
[30] "Intellectual Autobiography", 55.
[31] 70.
[32] See above p. 21 and note 7.
[33] "Are There Alternative Logics?" 90.
[34] *Probleme und Resultate, Vol. IV/1*, 9.
[35] See P. Lorenzen, "Protologic" for a reformulation of the program on the basis of a pragmatics of life-world argumentation see C. F. Gethmann, *Protologik*, 44–57.
[36] *Probleme und Resultate. Vol. IV/1*, 9–10.
[37] See esp. G. Gentzen, "Untersuchungen".
[38] There are no examples for contextual variant and party-invariant rules.
[39] See C. F. Gethmann, *Protologik*, sec. 2–4.
[40] For the proof of this 'inversion principle', already stated by G. Gentzen, see P. Schröder-Heister, "The Completeness".
[41] In this way one obtains rule schemes instead of the rules introduced so far. For a consequent reconstruction of rule scheme one has to add in C. F. Gethmann, *Protologik* (p. 157, following (4–4): "Für Propositionen mit Quantoren gilt folgende Schema-Schreibweise: kommt in einer Proposition eine gebundene Variable x vor, dann wird $\bigwedge xA$ bzw. $\bigvee xA$ notiert; kommt in einer Proposition A ein Parameter a vor, der für alle Vorkommnisse von x eingesetzt wird, dann wird A_a^x notiert." The expressions used *loc. cit.* 157ff have to be corrected accordingly.
[42] For questions of methodical connections of types of calculi see C. F. Gethmann, "Zur methodischen Ordnung regellogischer Kalkültypen".
[43] For details see C. F. Gethmann, "Zur formalen Pragmatik des Negators".
[44] See I. Johansson, "Der Minimalkalkül". C. F. Gethmann, *Protologik*, 177 Note 5 has to be specified correspondingly.
[45] See esp. P. Hinst, "Fundamentalsemantische Grundlegung".
[46] See e.g. P. Montague, R. C. Stalnaker. See the criticism in C. F. Gethmann, *Protologik*, 24–28, 62–63.
[47] There are similar approaches in the conception of "radical pragmatics" of F. Kambartel and in the conception of "fundamental semantics" of P. Hinst (see bibliography).
[48] *Probleme und Resultate. Vol. IV/1*, 7–8.
[49] *Loc. cit.* 24. (my translation)
[50] It should be noted that according to the examination of A. Kamlah the members of the Vienna Circle were not descriptivists (*Die normativ-analytische Wissenschaftstheorie*).
[51] *Loc. cit.* 23.
[52] *Loc. cit.* 24.

[53] In contrast to this Stegmüller makes a clear cut distinction between cognitive and social structures of science (*Probleme und Resultate. Vol. IV/1*, 16–17, 20).
[54] For more details see C. F. Gethmann, "Zur normativen Genese wissenschaftlicher Institutionen"; "Wissenschaftsforschung".
[55] R. Hegselmann and W. Raub, "Zur Logikabhängigkeit".
[56] See above p. 29.
[57] See D. Prawitz and R.-E. Malmnäs, "A Survey".

REFERENCES

Brouwer, Luitzen Egbertus Jan: "De onbetrouwbaarheid der logischen principes", *Tijdschrift voor Wijsbegeerte* **2** (1908), 152–158.

Brouwer, Luitzen Egbertus Jan: "Intuitionism and formalism", *Bull. Amer. Math. Soc.* **20** (1913), 81–96.

Brouwer, Luitzen Egbertus Jan: "Intuitionistische Mengenlehre", *Jahresbericht d. deutschen Math. Vereinigung* **28** (1920), 203–208.

Brouwer, Luitzen Egbertus Jan: "Über die Bedeutung des Satzes vom ausgeschlossenen Dritten in der Mathematik, insbesondere in der Funktionentheorie". *Journal für reine und angewandte Mathematik* **154** (1925), 1–8.

Carnap, Rudolf, Hahn, Hans, and Neurath, Otto: "Wissenschaftliche Weltauffassung: Der Wiener Kreis (The Scientific Conception of the World: The Vienna Circle)", in Otto Neurath, *Empiricism and Sociology* ((eds.) M. Neurath and R. S. Cohen), Dordrecht 1973, 299–318.

Carnap, Rudolf: *Abriß der Logistik*, Wien 1929.

Carnap, Rudolf: *The Logical Syntax of Language*, London 1967, 7th edition.

Carnap, Rudolf: *Introduction to Semantics*, Cambridge (Mass.) 1942.

Carnap, Rudolf: "Testability and Meaning", *Philosophy of Science* **3** (1936), 419–471; **4** (1937), 1–40.

Carnap, Rudolf: *Foundations of Logic and Mathematics*, Chicago 1963, 10th edition.

Carnap, Rudolf: "Intellectual Autobiography", in P. A. Schilpp (ed.), *The Philosophy of Rudolf Carnap*, La Salle (Ill.) 1963, 3–84.

Gentzen, Gerhard: "Untersuchungen über das logische Schließen", in M. E. Szabo (ed.), *The Collected Papers of Gerhard Gentzen*, Amsterdam 1969, 68–131.

Gethmann, Carl Friedrich: "Zur normativen Genese wissenschaftlicher Institutionen", in C. Burrichter (ed.), *Probleme der Wissenschaftsforschung*, Erlangen 1978, 69–91.

Gethmann, Carl Friedrich: *Protologik, Untersuchungen zur formalen Pragmatik von Begründungsdiskursen*, Frankfurt a.M. 1979.

Gethmann, Carl Friedrich (ed.): *Theorie des wissenschaftlichen Argumentierens*, Frankfurt a.M. 1980.

Gethmann, Carl Friedrich: "Wissenschaftsforschung? Zur philosophischen Kritik der nach-Kuhnschen Reflexionswissenschaften", in P. Janich (ed.), *Wissenschaftstheorie und Wissenschaftsforschung*, München 1981, 9–38.

Gethmann, Carl Friedrich (ed.): *Logik and Pragmatik. Zum Rechtfertigungsproblem logischer Sprachregeln*, Frankfurt a.M. 1982.

Gethmann, Carl Friedrich: "Zur methodischen Ordnung regellogischer Kalkültypen", in C. F. Gethmann (ed.), *Logik and Pragmatik*, 53–77.

Gethmann, Carl Friedrich: "Zur formalen Pragmatik des Negators", *Philosophica* **35** (1985), 39–67.

Hahn, Hans: "Logic, Mathematics and Knowledge of Nature" (transl. Arthur Pap), in A. J. Ayer (ed.), *Logical Positivism*, New York 1959, 147–161.

Hegselmann, Rainer and Raub, Werner: "Zur Logikabhängigkeit wissenschaftstheoretischer Paradoxien", *Erkenntnis* **17** (1982), 349–359.

Heyting, Arend: "Die formalen Regeln der intuitionistischen Logik", Sitzungsberichte der Preußischen Akademie der Wissenschaften, Physikalisch-Mathematische Klasse, Berlin 1930, 42–56.

Heyting, Arend: "Die formalen Regeln der intuitionistischen Mathematik", Sitzungsberichte der Preußischen Akademie der Wissenschaften, Physikalisch-Mathematische Klasse, Berlin 1930, 57–71; 158–169.

Hinst, Peter: "Fundamentalsemantische Grundlegung der Logik und strukturtheoretische Rekonstruktion der Interpretationssemantik", in J. Mittelstrass and M. Riedel (eds.), *Vernünftinges Denken. Studien zur praktischen Philosophie und Wissenschaftstheorie*, Berlin 1978, 52–70.

Hinst, Peter: "Pragmatische Regeln des logischen Argumentierens", in C. F. Gethmann (ed.), *Logik und Pragmatik*, 199–215.

Johannsson, Ingebrigt: "Der Minimalkalkül, ein reduzierter intuitionistischer Formalismus", *Compositio Mathematica* **4** (1937), 119–136.

Kambartel, Friedrich: "Überlegungen zum pragmatischen und argumentativen Fundament der Logik", in K. Lorenz (ed.), *Konstruktionen versus Positionen*, Bd. 1, Berlin 1979, 216–228.

Kambartel, Friedrich: "Pragmatische Grundlagen der Semantik", in C. F. Gethmann (ed.), *Theorie des wissenschaftlichen Argumentierens*, 95–114.

Kambartel, Friedrich: "The Pragmatic Understanding of Language and the Argumentative Function of Logic", in H. Parret and J. Bouveresse, *Meaning and Understanding*, Berlin 1981, 402–437.

Kambartel, Friedrich: "Formalistische und sophistische Elemente in der 'dialogischen Logik'", in C. F. Gethmann (ed.), *Logik und Pragmatik*, 41–52.

Kambartel, Friedrich and Schneider, Hans-Julius: "Constructing a Pragmatic Foundation for Semantics", in Floistadt and Guttorm (eds.), *Contemporary Philosophy. A New Survey* **1**, The Hague 1981, 155–178.

Kamlah, Andreas: *Die normativ-analytische Wissenschaftstheorie der logischen Analyse*, Mannheim 1979 (eds. H. Albert, M. Küttner, and H. Lenk, Beiträge zur Philosophie und Methodologie der Einzelwissenschaften).

Kaufmann, Felix: *The Infinite in Mathematics. Logico-Mathematical Writings* (ed. by B. McGuinness), Dordrecht 1978.

Kutschera, Franz von and Breitkopf, Alfred: *Einführung in die moderne Logik*, Freiburg 1971 (1979) 2nd edition.

Lenzen, Wolfgang: *Theorien der Bestätigung wissenschaftlicher Hypothesen*, Stuttgart-Bad Cannstatt 1974.

Lorenzen, Paul: "Protologik. Ein Beitrag zum Begründungsproblem der Logik", *Methodisches Denken*, Frankfurt a.M. 1968, 81–93.

Lorenzen, Paul and Schwemmer, Oswald: *Konstruktive Logik, Ethik und Wissenschaftstheorie*, Mannheim 1975.

Neurath, Otto: "Ways of the Scientific World-Conception", in Otto Neurath, *Philosophical Papers 1913–1946* (eds. R S. Cohen and M. Neurath), Dordrecht 1963.

Popper, Karl R.: "Why are the Calculi of Logic and Arithmetic Applicable to Reality", *Conjectures and Refutations*, London 1963, 201–214.
Prawitz, Dag and Malmnäs, P.-E.: "A Survey of Some Connections between Classical, Intuitionistic and Minimal Logic", in eds. H. A. Schmidt, K. Schütte, and H. G. Thiele, *Contributions to Mathematical Logic.* Proceedings of the Logic Colloquium. Hannover 1966, Amsterdam 1968, 215–229.
Prior, Arthur N.: "The Paradoxes of Derived Obligation", *Mind* **63** (1954), 64–65.
Prior, Arthur N.: *Formal Logic*, Oxford 1962, 2nd edition.
Quine, Willard Van Orman: *Philosophy of Logic*, Englewood Cliffs 1970, 3rd edition.
Schleichert, Hubert (ed.), *Logischer Empirismus–Der Wiener Kreis*, München 1975.
Schroeder-Heister, Peter: "The Completeness of Intuitionistic Logic with Respect to a Validity Concept Based on an Inversion Principle", *Journal of Philosophical Logic* **12** (1983), 359–377.
Stegmüller, Wolfgang: *Das Wahrheitsproblem und die Idee der Semantik*, Wien 1968, 2nd edition.
Stegmüller, Wolfgang: *Probleme und Resultate der Wissenschaftstheorie und Analytischen Philosophie. I: Wissenschaftliche Erklärung und Begründung*, Berlin 1969.
Stegmüller, Wolfgang: *Probleme und Resultate der Wissenschaftstheorie und Analytischen Philosophie. II: Theorie und Erfahrung*, Berlin 1970.
Stegmüller, Wolfgang: "Neue Betrachtungen über die Ziele und Aufgaben der Wissenschaftstheorie", *Probleme und Resultate der Wissenschaftstheorie und Analytischen Philosophie. IV/I: Personelle und statistische Wahrscheinlichkeit*, Berlin 1973, 1–64.
Stegmüller, Wolfgang: *Hauptströmungen der Gegenwartsphilosophie*, 2, Stuttgart 1975, 2nd edition.
Tetens, Holm: *Bewegungsformen und ihre Realisierungen. Wissenschaftstheoretische Untersuchungen zu einer technikorientierten Rekonstruktion der klassischen Mechanik*, Erlangen (phil. Diss.) 1977.
Toulmin, Stephen: "From Logical Systems to Conceptual Populations", *Boston Studies in the Philosophy of Science* **8** (1971), 552–564.
Toulmin, Stephen: *Human Understanding. Vol. I: General Introduction and Part I*, Oxford 1972.
Waismann, Friedrich: "Are There Alternative Logics?", *How I see Philosophy*, London 1978, 67–90.
Whitehead, Alfred N. and Russell, Bertrand: *Principia Mathematica*, Bde. 1–3, Cambridge 1910.
Wittgenstein, Ludwig: *Tractatus logico-philosophicus.* (transl. D. F. Pears and B. F. McGuinness), London 1961.
Wittgenstein, Ludwig: *Philosophical Investigations*, eds. G. E. M. Anscombe, R. Rhees, G. H. v. Wright, Oxford 1956.

FRIEDRICH KAMBARTEL

THE PRAGMATIC UNDERSTANDING OF LANGUAGE AND THE ARGUMENTATIVE FUNCTION OF LOGIC

Pragmatic understanding of language means that language structures, especially logical structures, are reconstructed throughout not as *objects* but as part of a rational *practice*.[1] Such a pragmatic reconstruction cannot draw on those conceptions of language and logic that at present are dominant in formal logic and linguistics. These are mostly based on the so called "*semantic* point of view" which, in its modern form, goes back to Tarski. Here one has to take the word 'semantic' not just in the general Greek sense indicating only that the meaning of language expressions is treated. Rather, the word 'semantic' refers to a quite peculiar theory of meaning. This theory assumes that language consists of special (structured) objects, which become meaningful by their being related to certain other objects so to speak "in the world". Among these other objects we have to imagine such abstract objects as functions, especially truth functions. These meaning relations are introduced and analysed by using another language (or level of language) which is usually called "(the) metalanguage". In the metalanguage, so we are informed, we can speak about the object language, i.e. that language which is the 'object' of our considerations.

As an illustration let us consider the following example. In logic books written from a semantic point of view, one can find truth-conditional definitions of the universal quantifier of the following kind: '$\bigwedge_x a(x)$' is by definition true if and only if for all "admitted" substitutions x/n the resulting $a(n)$ is true. – More sophisticated versions using the possible worlds concept go roughly like this: A possible world w belongs to the meaning of a universal sentence '$\bigwedge_x a(x)$' if and only if for all substitutions x/n: w is an element of the meaning of $a(n)$.

Strictly speaking, from the semantic viewpoint we characterize with such definitions abstract objects like sets or functions and coordinate them as meanings with certain other objects, called language expressions or symbols. Quite obviously this type of explaining universal quantification has serious deficiencies: Besides the unjustified ontological framework it already uses expressions like 'for all x'. Therefore

Robert E. Butts and James Robert Brown (eds.), Constructivism and Science, 47–57.

it is not helpful for *learning* or *understanding* universal quantification, contrary to what we would expect from logic books. This deficiency is not avoided, as most logicians since Tarski think, by just taking '$\bigwedge_x$' as a symbol of the object language and 'for all *x*' as an expression of English, taken as a metalanguage to the language of formal logic. For if one wants to *understand* the use of universal quantifiers, it would be a *petitio principii* to simply use metalinguistic translations of (or equivalents to) those terms of the object language that we want to introduce.

Also it would be a (more indirect) *petitio principii*, if such a translation or equivalent is used to build up the metalanguage level. Nevertheless methodical circles of this kind are characteristic of the semantic point of view. Not taking them seriously means giving up the goal of understanding of what we do when we use language, especially when we use logical expressions and symbols.

So semantical analysis of the Tarski-type does not provide us with a rational understanding of meaning, neither of the lexical nor of the categorical component of meaning. Semantics in this sense may give us a very general description of language structures, a description though, which *works* only on the presupposition that we have already acquired all the relevant lexical and categorical competence. Hence this sort of semantics has nothing to do with the intensional part of language analysis – if we still want to use the words 'Sinn' and 'intension' in the Frege–Carnap tradition.

But couldn't we take the pragmatic approach of Austin, Searle and others as an alternative? Linguistic pragmatism, in its general formulations, surely points to a trivial truth, namely that language is a system of *acts*, not objects. Unfortunately in its concrete performance the "pragmatic" approach often is only *pseudo*pragmatic, namely in those cases where pragmatic linguists just take the traditional semantic or logico-grammatic categories and "invent" speech acts on the verbal level, as in the case of "propositional acts" derived from the stipulated abstract objects called propositions. Thus pragmatics here depends on non-pragmatic distinctions and an objectifying surface-understanding of language. Therefore this kind of pragmatics still gives us no knowledge of how we can learn speech acts without having at our disposal a semantic conception of language. It gives us no knowledge, e.g. of how we can *learn* a propositional act *a* without just defining it as "expressing the proposition *a*". Let me therefore propose a radicalization of the pragmatic approach, which (re)constructs language as a rational system of acts without "semantic" rests at its basis and which I like to call *constructive pragmatics*.[2]

Speech acts are acts which we do (actualize) by doing (actualizing) other acts, like putting on the light by operating a switch. I propose to call an act *mediated* by other acts if and only if this act, by its definition, can be actualized by actualizations of one or several of these other acts. We may speak of *mediated acts* (absolutely) if and only if an act can be performed in a mediated way only.

Let us now turn directly to those mediated acts which constitute language and other human institutions. I shall call them *symbolic acts*. Symbolic acts serve the purpose of agreeing about the situation in which we are, and to cooperate (pragmatically) by influencing other people's actions. They, in a sense, make rational life possible. We construct (constitute) symbolic acts by giving a special (symbolic) use to other acts. This is done by *agreements* which restrict (explicitly or tacitly) the actualization of these acts to certain situations or lay down for them certain consequences in the course of our actions. Think e.g. of the phonetic act "come" (or rather [k ʌ m]) which by agreement has to be understood as a specific demand. A person who has uttered a demand must – by the sort of agreements which are characteristic for demands – be prepared to see his demand fulfilled by the addressee (no sanctions however implied, because *demands* are not necessarily *commands*).

The performance of symbolic acts is built on the possibility of following rules. However we should not confuse symbolic action with following a rule, e.g. a rule that an act a is permitted only in certain situations or is required in certain other situations. A rule $R(a)$ just concerns a regular way of doing or forbearing a. That is: we have a different rule, if we regulate a different a' in the same way R. Drinking tea every morning is different from drinking coffee every morning. Whereas in the case of a symbolic act the a underlying the regulation R does not really matter: it is methodically arbitrary, though there may be historical and mnemotechnical reasons for a particular choice. Any a' about which we make the same agreement (R) as about a, in principle will be good for the same symbolic intention, though in fact naturally we do not and cannot make all possible agreements. Thus actualizing the symbolic act belonging to the rule or rule system R means to perform *one of the* acts, which by agreement are given the same symbolic use. This illustrates the particular way in which agreements and mere conventions are combined in symbolic acts.

As is generally the case with mediated acts, we follow at least two pragmatic intentions in the performance of a symbolic act, the *direct* one of the mediating act and the *indirect* one of the symbolic act. We shall say that the direct intention 'carries' the symbolic one.

A symbolic act *S* has *informative* meaning with reference to the conditions which by agreement make the performance of *S* a correct one, and it has *performative* meaning in so far as it has constitutive pragmatic consequences. Obviously mixed cases are possible.[3]

As you will know by your own practice, in *elementary* speech situations we can learn an *elementary* symbolic act *S* by exemplification and we can *pragmatically* control our understanding of the agreement which is intended. We just get acquainted with examples and counter-examples of a correct actualization of *S* in an appropriate pragmatic context. And in order to control and correct our understanding we can – in elementary situations – always return to what we do, especially what we do in non-symbolic action. (Take for an example the demand "come!" mentioned above.) Therefore in the elementary case, in order to acquire a symbolic act, we, in principle at least, need no description using other symbolic acts. So there is no metalanguage-circularity.

In natural language, if we do not have an exact knowledge of the situation, it is sometimes not clear what kind of elementary symbolic acts are performed. E.g. uttering "bricks" may carry symbolic intentions like:

1. Bricks are over there.
2. Get me some bricks!
3. Bricks are over there, get me some!

Or in a sufficiently complex situation we may not know whether a warning 'hot' concerns the soup or the pan. In such cases, if we want to make clear what we have in mind, it is often helpful just to connect the corresponding symbolic act *Q* with a further symbolic act *P*. So in our example we could say 'soup hot' or 'pan hot'. Naturally a suitable *P* must be already at our disposal, i.e. there must be a well known use of *P* in the considered language community. Let us in the described speech situation call *P* a *specification* of *Q*, and let us write specifications in the form $P-Q$. In natural English or German we may use quite a variety of grammatical structures for specifications, e.g. the statement form '*P* is *Q*' (with emphasis on *P*) or adverbial constructions (like 'hot soup').[4]

Remaining disambiguities might be dissolved by using additional deictic or indexical expressions like "this", "that", "here", "there", and, naturally, the personal pronouns. But all this clearly does not lead us to "invariant" statements about the situation (as opposed to "variant" statements in the sense of utterances having situation-relatively changing symbolic intentions), statements e.g. like "Peter is small".

I propose to reconstruct invariant elementary statements as follows: Let *P*, *Q* be informative elementary symbolic acts. Then in many relevant cases we may agree upon rules of the following type:

$P \Rightarrow P-Q$
(read: if *P* is correct, then *Q* (specified by *P*) is correct, or misleadingly: *P*-situations are *Q*-situations)

Examples are:

Peter $\Rightarrow$ tall	(Peter is tall)
horse $\Rightarrow$ animal	(horses are animals)

Further elementary statements are the negative counterparts

$P \not\Rightarrow P-Q$
(read: if *P* is correct, then *Q* (specified by *P*) is incorrect or misleadingly: *P*-situations are not *Q*-situations).

What is the use of invariant statements "*P* is *Q*" in our life? Well, wherever an elementary statement "*P* is *Q*" is correct ("valid"), there in a *P*-situation (that is: a situation, where *P* can be correctly performed) you have the possibilities to act which are characteristic for a *P*–*Q*-situation. This means that, e.g. in the case of a correctly done elementary statement "Peter is small" we take it that in situations where "Peter" is correct (objectively spoken: where Peter is involved), we may rely on the situation allowing a correct performance of "small" (in connection with "Peter"). Normally this holds only for a certain range of Peter-situations, which is different, e.g. for the usual sense of "Peter is small" and "Peter is lecturing". By invariant elementary statements we thus take the first step towards agreements treating situations "generally", "in an abstract way". This enables us to go, via language, beyond the present, concrete situation.

Elementary statements may by agreement come to carry a further symbolic intention and thereby be transformed into elementary *assertions*. The need for elementary assertions arises in situations, where action already depends on correctly performed elementary statements, and where the participants do not agree on the correctness of such a performance. In this case one can either give up common orientation as provided by elementary statements, or try to overcome private opinions at a new level of transsubjectivity, namely *argumentation*. Argumentation means here quite simply all attempts to settle differences on the basis of previously or newly established agreements. Thus we may go back to other elementary statements, which are not at issue, or reproduce (resp. once more realize) the original learning situation of the words involved. In the first case, e.g. if we have already agreed upon '*P* is *Q*' and '*Q* is *R*', this might stop further discussion on '*P* is *R*'. Often it will be helpful (or even necessary) to remember the aims and activities which originally gave rise to the terms of a disputed elementary statement. This may be sufficient to indicate what activities we can enter into for the purpose of justifying an elementary statement by arguments. Someone who now not only just states something, but *asserts* what he is stating, must be prepared to establish by argumentation a transsubjective agreement that his statement has been made correctly. In this way statements characteristically come to carry a further "assertive" symbolic intention, and then are called *assertions*.

We shall now consider logically *complex assertions* and show that, once we have elementary assertions, it is helpful to proceed to more complex activities related to the logical words of our language.

In planning common practice, it is often advisable to make action depend on more than one justified orientation. Thus it may be that one has to do two things, justify *a* and justify *b*, to make someone else enter into a certain activity. Whoever thinks himself capable of fulfilling both these justification-tasks, claims that he can do a complex justification work, as we might call it. This claim is raised in English in the form of a complex statement, the *conjunction* '*a* and *b*' in logical symbols: '$a \wedge b$'. But is not there still a *petitio principii* in this reconstruction of logical conjunction? Did we not use the word 'and', when we put the relevant complex justification task as follows: 'to do two things, justify *a and* justify *b*'? Indeed we did. But here (in this formulation) the word 'and' is not a logical connective (junctor), but a word which we use to form expressions for complex *acts*, e.g. in the case of commands. And what it means to perform an act by doing several (in our case two) other acts we can learn in practice by suitable examples. This practical learning does *not* imply that we need to *speak about* this learning situation and use, in doing this, logical conjunction. Thus there is no circle of definition.

In a similar way we obtain logical *adjunction* '*a or* (*vel*) *b*', in symbols: '$a \vee b$'. Adjunction serves among other ends the purpose of planning action which is dependent on several possibilites (alternatives). Again we first have to establish a pragmatic basis for logic, namely learn a sort of pragmatic complexity, which we may express as doing one of two things (acts). Knowing this we know in particular what it means to fulfil one of the (two) justification-tasks related to assertions *a*, *b*. And this new justification-claim we may again attach to a new complex statement '*a or b*' ('$a \vee b$').

The purpose, e.g. of dividing up justifications into parts, leads us to logical *subjunction* '*b, if a*' ('$a \to b$'). In this case the constitutive claim is: let me have a justification of *a*, then I shall be able to construct one for *b*. This should, for invariance reasons, include the case that I have an independent justification of *b*.

As to the negator 'not' ('$\neg$'): whoever puts forward a *negation* $\neg a$ (*not a*), claims to have a procedure, by which we can be sure that attempts to justify *a* will fail. In short: the proponent of a negation $\neg a$ must be able to *refute a*. Obviously negations provide us with an important kind of knowledge, because with them we can, e.g. give up trying to justify the negated *a* or reflecting on courses of action presupposing the validity of *a*. Again negation makes use of a pragmatic distinction, namely between succeeding in and failing at doing something (acting in a certain way).

We are now in a position to settle the question of *tertium non datur*: Is $a \vee \neg a$ valid independently of the content of *a*, i.e. valid by taking into account nothing else but the meaning of the logical words $\vee$, $\neg$? Obviously not, because there are *a*'s for which we have neither justifications nor refutations at hand.

The *principle of contradiction* on the other hand holds, i.e. $\neg(a \wedge \neg a)$ is logically valid. This is so, because nobody can do both justify an assertion *a*, and at the same time show us that an attempt to justify *a* must fail.

Logical quantifiers may be treated similarly: Asserting 'for all *x*: *a*(*x*)', in symbols: '$\bigwedge_x a(x)$' (*universal quantification*), means having a *procedure* at hand to justify $a(n)$ for an arbitrary substitution x/n. To know a procedure does not necessarily imply descriptions using universal quantifiers: to know how to get nails into a concrete wall does not imply understanding *sentences* like 'all nails are in the wall' or 'for all nails I know how to get them into the wall'.

Who asserts 'there is an x, such that: $a(x)$', in symbols: '$\bigvee_x a(x)$' (*existential quantification*), claims to have a substitution x/n at hand, so that he can give a justification for the corresponding $a(n)$. Existential quantification in this sense is "effective": you have to know *how to "construct"* the substitution which is necessary to defend the existential assertion. Obviously no logical platonism or axiomatic formalism is necessary to understand the meaning of the logical words and symbols and to judge on logical validity. What one has to do is simply to reconstruct their rational pragmatic place in our lives, i.e. one has to understand them as part of rational action, namely in this case of argumentation.

Obviously the proposed reconstructions of complex assertions are related to the *dialogical logic* of P. Lorenzen and K. Lorenz.[5] The dialogical approach too claims to characterize, by schematic rules, the logical aspect of concrete argumentation as exemplified e.g. by socratic dialogues. And yet the short history of dialogical logic shows that the latter has not yet managed to fully overcome the distortions caused by the mathematization of logic.

Already the first version of Lorenzen[6] stylizes argumentation as following formal rules in a dialogical 'game', analogous to the moves in chess. Lorenz[7] then has worked out this game-theoretical reconstruction of formal obligations in argumentation and of logical validity. He gave precise and schematically applicable formulations not only for those steps which depend on the *logical particles* (*particle rules*), but also for the *general frame* which has to be observed when making any particle-related moves (*frame rules*). For all these rules, especially for frame rules, we can raise the question of justification, the question, that is, whether or not they are an adequate reconstruction of logical complexity in concrete argumentation. And here the adequacy control can lead to a reasonable regulation only if we already understand the rational place of logically complex argumentation in our lives.[8]

The above proposals for an argumentative use of logical words are meant to offer such an understanding. Classically spoken, they play the role of a semantics in relation to the 'monological' *or 'dialogical'* formal systems.

Lorenzen and Lorenz on the other hand transform the adequacy problem of schematical argumentation rules into a completeness problem which can be treated on the *formal level* alone and thus is located again inside mathematical logic. They take the 'dialogical' games themselves as an argumentative semantics for logical calculi, i.e. they conceive of these games as giving meaning to the logical words. In that case (after this transformation) one can naturally discuss whether the derivability of a logical formula in a calculus is equivalent to the existence of a winning-strategy in a two-persons-game. But the unfinished discussion about the justification or rationality of the dialogical rules, especially the frame rules[9] shows that there is still an element of arbitrariness in dialogical logic, which cannot be removed on the technical level alone. At the beginning Lorenzen was still aware of the connection of agonal ('eristic') or cooperative ('dialectical') argumentations with different schematical regulations.[10] Later, for unclear reasons, the *agonal* version was formally favoured, and in the discussion of the frame rules technical aims like, e.g. 'simplicity' or mere plausibility considerations prevail.[11]

Let us look back on the systematic course, especially the method, which we have followed. We may call our considerations and proposals critical reconstructions of an activity, which we all know in use, namely language. Critical reconstructions constitute an understanding of human practice, which is more than a mere surface description. Reconstructions being *constructions*, let me first dwell upon this point. 'Construction' in my use is a term of the *theory of action*. In a (pragmatic) *construction* we build up, step by step, systematically connected actions. Pragmatic constructions are *justified*, in so far as they lay the basis for a rational life. To intend justified constructions implies that, besides the question of how we can acquire step-by-step certain acts, we have to consider the connection between our acts and our aims: That is to say we have to judge the consequences of performing those acts, i.e. evaluate the resulting situations. In this sense justified *pragmatic* constructions include *teleological* considerations.

In most cases of pragmatic constructions, we do not enter into the corresponding actions for the first time. Rather we are already involved in them in a confused and sometimes seriously distorted way. Thus in some sense we always know what we are doing, and in some sense we do not know; i.e. we know how to do certain things without having a clear awareness either of the pragmatic construction of this ability, or of its teleological place in our life. This leads us to misunderstand our activities and sometimes thereby to become subject to a misuse of our life without our realizing it. Therefore in such situations it is advisable to work out clear pragmatic and teleological constructions for the actions we are involved in, in short: to *reconstruct* what we are doing. I think that this belongs to striving for *enlightenment*; and obviously the actual use of *language* often is confused enough, by itself and moreover by logicians and linguistics, to need this enlightenment.

NOTES

[1] This essay includes material that was presented at the Conference *Semantics from Different Points of View* (published in: Kambartel, F., "Constructive Pragmatics and Semantics", in Bäuerle, R., Egli, U. and von Stechow, A. (eds.), *Semantics from Different Points of View*. Berlin-Heidelberg-New York: Springer, 1979, 196–205). The present paper is a contribution to the Round Table discussion on *Discourse and the Rational Reconstruction of Logic*, held at the Cerisy Conference on *Meaning and Understanding*. I argued at this Round Table that the reconstruction of meaning is based completely on distinctions which belong to a theory of action and interaction, involving neither mentalistic nor ontological assumptions. Logical argumentation, as well as the methodically precedent elementary levels of stating or asserting something, can be successfully analysed in this way. This foundation of logical meaning is very similar to Kolmogorov's logical theory of task solving (Kolmogorov, A., "Zur Deutung der intuitionistischen Logik", *Mathematische Zeitschrift* **35**, 1932, 58–65; cf. Heyting, A., *Mathematische Grundlagenforschung – Intuitionismus, Beweistheorie*. Berlin-Heidelberg-New York: Springer, 1934 (1974), 14sqq.) and can serve as a justification for the constructive and the intutionistic approach to logic.

[2] For a broad orientation about the constructive approach in the philosophy of language and theory of science cf. Lorenzen, P., *Normative Logic and Ethics*. Mannheim-Zürich: Bibliographisches Institut, 1969; Kamlah, W. and Lorenzen, P., *Logische Propädeutik – Vorschule des vernünftigen Redens*. Mannheim: Bibliographisches Institut, 1967 (1973); Lorenzen, P. and Schwemmer, O., *Konstruktive Logik, Ethik und Wissenschaftstheorie*. Mannheim-Zürich: Bibliographisches Institut, 1973 (1975), and Kambartel, F., *Theorie und Begründung. Untersuchungen zum Philosophie- und Wissenschaftsverständnis*. Frankfurt a.M.: Suhrkamp, 1976. For a pragmatic reconstruction of understanding and meaning as proposed in this essay, cf. Kambartel, F., "Symbolic Acts – Remarks on the Foundation of a Pragmatic Theory of Language", in Ryle, G. (ed.), *Contemporary Aspects of Philosophy*. Stockfield-London-Boston: Oriel Press, 1977, 70–85; Schneider, H., *Pragmatik als Basis von Semantik und Syntax*. Frankfurt a.M.: Suhrkamp, 1975, Lorenz, K., *Elemente der Sprachkritik – Eine Alternative zum Dogmatismus und Skeptizismus in der analytischen Philosophie*. Frankfurt a.M.: Suhrkamp, 1970; Lorenz, K., "Sprachtheorie als Teil einer Handlungstheorie. Ein Beitrag zur Einführung linguistischer Grundbegriffe", in Wunderlich, D. (ed.), *Wissenschaftstheorie der Linguistik*.

Kronberg: Atheneaum, 1976; Mittelstraß, J., "Das normative Fundament der Sprache", in Mittelstraß, J. (ed.), *Die Möglichkeit der Wissenschaft.* Frankfurt a. M.: Suhrkamp, 1974, 158–205. For literature concerning constructive logic see below, Notes 5, 6.

[3] For more detailed information about the preceding distinctions and their applications, cf. Kambartel, F., 1977 (cf. Note 2); Kambartel, F., "Pragmatische Gundlagen der Semantik", in Gethmann, C. (ed.), *Theorie des wissenschaftlichen Argumentierens.* Frankfurt a.M.: Suhrkamp, 1980, 95–114.

[4] For the analysis of the first level of symbolic complexity I follow Schneider, H., "Sprachtheorie auf pragmatischer Gundlage", in *Arbeitsgruppe Semiotik* (eds.), *Die Einheit der semiotischen Dimensionen.* Tübingen: Gunter Narr, 1978, 171–189, especially 184sqq.

[5] The development of the 'dialogical' foundation of logic is now documented in the collection Lorenzen, P. and Lorenz, K., *Dialogische Logik.* Darmstadt: Wissenschaftliche Buchgesellschaft, 1978. Cf. also Lorenzen, P., 1969 (cf. Note 2); Lorenz, K., "Rules versus Theorems – A new Approach for Mediation between Intuitionistic and Two-Valued Logic", *Journal of Philosophical Logic 2*, 1973, 352–369.

[6] Lorenzen, P., "Logik und Agon", *Atti del XII Congresso Internazionale di Filosofia, IV.* Firenze, 1960, 187–194. Reprinted in Lorenzen, P. and Lorenz, K., 1978, 1–8 (cf. Note 5).

[7] Lorenz, K., *Arithmetik und Logik als Sprachspiele.* Diss. Universität Kiel, partly reprinted in Lorenzen, P. and Lorenz, K., 1978, 17–95 (cf. Note 5); Lorenz, K., "Dialogspiele als semantische Gundlage von Logikkalkülen", *Archiv für mathematische Logik und Grundlagenforschung* **11**, 1968, 32–55, 73–100, reprinted in Lorenzen, P. and Lorenz, K., 1978, 96–162 (cf. Note 5).

[8] This perspective has been worked out to some degree in Berk, U., *Konstruktive Argumentationstheorie.* Stuttgart-Bad Canstatt: Fromann-Holzboog, 1979; Kambartel, F., "Überlegungen zum pragmatischen und zum argumentativen Fundament der Logik", in Lorenz, K. (ed.), *Konstruktionen versus Positionen*, I. Berlin: De Gruyter, 1979, 95–114; Gethmann, C., "Die Ausdifferenzierung der Logik aus der vorwissenschaftlichen Begründungs- und Rechtfertigungspraxis", *Zeitschrift für Katholische Theologie*, 102, 1980, 24–32; Gethmann, C., *Protologik. Untersuchungen zur formalen Pragmatik von Begründungsdiskursen.* Frankfurt a. M.: Suhrkamp, 1980. Already Lorenz, K. in "Die Ethik der Logik" (in Gadamer, H. G. (ed.), *Das Problem der Sprache.* München: Fink, 1967, 81–86) took some steps in this direction.

[9] Cf. e.g. Lenk, H., "Philosophische Logikbegründung und rationaler Kritizismus", in Lenk, H. (ed.), *Metalogik und Sprachanalyse.* Freiburg: Rombach, 1973, 88–109; Stegmüller, W., "Remarks on the Completeness of Logical Systems relative to the Validity Concept of P. Lorenzen and K. Lorenz", *Notre Dame Journal of Formal Logic* **5**, 1964, 81–112. But see also the permanent change in the arguments for the 'frame rules' (and correspondingly their form) in the developing proposals of Lorenz and Lorenzen.

[10] Cf. the final remarks in Lorenzen, P., 1960 (cf. Note 6).

[11] Cf. e.g. Lorenzen, P. and Schwemmer, O., 1973, 68 (cf. Note 2); Lorenz, K., 1968, 38 and 40 (cf. Note 7); Lorenzen, P. and Lorenz, K., 1978, 104sq. and 107 (cf. Note 5).

KUNO LORENZ

RULES VERSUS THEOREMS

A New Approach for Mediation between Intuitionistic and Two-Valued Logic

I

Contemporary critics of two-valued logic concentrate on the reasons for accepting the *tertium non datur* $A \vee \neg A$ as a valid propositional schema. Brouwer explicitly states[1] that only by unjustified extrapolation of logical principles from those which correctly describe the general relations among propositions on finite domains to those that allegedly regulate propositions on infinite domains, could it happen that $A \vee \neg A$ is accepted as valid. He was the first to observe that value-definite (decidably true or false) propositions do not generally transfer value-definiteness to their logical compounds. No better support could be found for the claim that the classical characterization of propositions as entities that are either true or false is inadequate. The union of the class of all true propositions and the class of all false propositions does not contain all logical compounds out of either true or false propositions; it does not contain, for example, certain as yet neither proven nor disproven universal propositions of elementary arithmetic. But nobody has seriously advanced the thesis that such propositions should not count as propositions at all.[2] In fact, it is generally conceded that the usual way to form finite and infinite logical compounds makes sense even if nothing can be said about their truth-value. It is obligatory, then, to look for a better introduction of the term 'proposition' than the classical one and, of course, not only a syntactical introduction, which is trivial, but a semantical one. The validity concept of two-valued logic being dependent on the value-definiteness of propositions will consequently have to be given up and replaced by a concept of validity that works without recourse to the truth-value of the propositions in question. It follows that the classical introduction of logical particles by the (finite or infinite) truth table method has to be given up as well, or rather it has to be amended in such a way that the general definition of logical particles restricted to value-definite propositions yield the well-known classical ones.

Robert E. Butts and James Robert Brown (eds.), Constructivism and Science, 59—76.

Various attempts in this direction have been made, most prominent, among others, the proof-theoretic interpretation of intuitionistic logic by Kolmogorov and the operative interpretation by Lorenzen.[3] In these attempts, the introduction of the junctor 'if – then' marks the starting point of a deviation from classical procedures. Kolmogoroff replaces the concept of truth by the concept of provability and the provability of $A \rightarrow B$ correspondingly by the provability of B relative to a proof of A that is hypothetically assumed. The concept of proof and *a fortiori* of provability has to be taken over from existing unformalized mathematics. Lorenzen replaces the concept of truth by the concept of (generalized) derivability within some calculus K, such that derivability of $A \rightarrow B$ has to be read as admissibility of the corresponding rule $a \Rightarrow \beta$ (with $A \leftrightharpoons \vdash_K a$ and $B \leftrightharpoons \vdash_K \beta$) relative to K. Here, the concept of admissibility has to be accepted as intuitively clear. Actual difficulties of interpretation occur in both cases after iterating the logical composition, e.g., with 'if – then', and no way out is visible if propositions other than mathematical ones are candidates for logical composition. Yet, these attemps have cleared the way to the additional insight that not only is value-definiteness not hereditary generally to logical compounds, but that it is possible to ascertain the (non-logical) truth of logically compound propositions without recourse to the truth-value of the subpropositions. There are, for example, true (i.e., provable resp. derivable) and not logically true subjunctions $A \rightarrow B$ without any knowledge even about the value-definiteness of A or B: one may take a suitably chosen admissible rule $a \Rightarrow \beta$ within an undecidable calculus.

A successful criticism of two-valued logic has to be able, therefore, to balance a wider concept of proposition with a correspondingly wider concept of logical composition and to add an adequate concept of validity for propositional schemata, or, alternatively, an adequate concept of (logical) implication by keeping the meta-equivalence 'A implies B' if and only if '$A \rightarrow B \in$ valid'. The usual method of devising formal systems, i.e., of introducing a syntactical concept of validity for syntactically defined well formed formulas is, of course, insufficient. Heyting's formalization of intuitionistic logic did permit precise comparison with other calculi, calculi of classical logic as well as of modal logic and others,[4] but it could not answer the initial question of what kind of theory or rather 'action' (*Denktätigkeit*)[5] it actually is that is formalized by this or that formal system of logic. And intuitionism (in the spirit of Brouwer) consequently never claimed to be able to represent its logic fully by a formal system.

In order now to gain a better understanding of the actual conflict, it is necessary to go beyond these introductory remarks by stressing a difference of points of view between the proponents of classical logic and the proponents of effective logic, which is so much taken for granted that it is hardly ever explicitly disputed. Since the time of Leibniz, classical logic is often referred to as a system of 'truths' which hold 'universally', in 'all possible worlds', and therefore independently of the special facts of the 'actual world', i.e. of the natural sciences. And mathematics is, following the programme of logicism, to be constructed as a special part of this system of logical truths. In precisely this sense logic, and with it mathematics, came to be considered as a system of tautologies without factual content. Classical logic is the formal frame for any scientific investigation, the a priori basis of empirical science. It makes no essential difference when Quine, expanding ideas of C.I. Lewis, disputes the distinction a priori-empirical and stresses the uniformity of the whole corpus of scientific truths instead.[6] This corpus is not uniquely determined by observational facts, it is in need of conceptual and other theoretical constructions, e.g. mathematical ones, which are chosen by intrinsic criteria of perspicuity, economy, connectedness *et alii* of the system of science as a whole. Thus, it may well obtain that even logic, one of the central parts of the corpus of truths, has to be changed due to new observational facts in order to satisfy the aforementioned criteria. Yet, even then, logic, the system of accepted logical truths, be it in its formalized version derivable by a classical or by some other calculus, remains the formal frame of science. Logic may be called 'relatively universal', i.e., a system of universal truths relative to the actual state of science.

On the other hand, effective logic (explicitly in its operative interpretation), must be looked at as a system of 'universal' rules which are accepted whenever a system of rules of action, e.g. rules for producing proofs or rules for producing arbitrary strings of signs, has been laid down. In this case, the field of application for the rules of logic is not the world as the totality of facts, but rather the world as seen in terms of specific kinds of scientific human activities. Within mathematics, for example, the rules of effective logic may be used without restriction. And this obtains because mathematics is not viewed as a system of truths, even less logical ones, but is treated as an independent scientific activity which, together with its intrinsic rules, may use the rules of logic as additional 'admissible' ones. Effective logic is the material content of any scientific investigation, the 'empirical core' even within mathematics. It goes without saying that no uniqueness claim is added. The set of rules of effective logic may vary from one scientific activity to another, and are not even strictly determined by any one of those. It coincides even with the set of rules of classical logic in the case of strictly finite mathematics, as Brouwer explicitly observed.[7] Thus, logic may again be called 'relatively universal', i.e., a system of accepted universal rules relative to the field of investigation.

In the light of these considerations, the basic conflict is a question rather of the set-up of formal logic itself than of accepting this or that propositional schema as valid. And, indeed, the competing views, to treat logic either as a set of rules ('for correct thinking') or as a set of theorems ('on the general behaviour of thought'), trace back to the very beginning of formal logic, to Aristotle and his interpretation by posterity. The conflict is known under the rubric: logic – art or science?, the respective Greek terms being '*τέχνη*' and '*ἐπιστήμη*'.

The problem, at the beginning of logic in the Greek period, was to set up a discipline that realizes the possibility of well-founded argumentation without using these very means of argumentation under pain of begging the question.

If logic, and in the case of Aristotle this means his syllogistic, would have to count as a science, it should obey the conditions laid upon a system of truths by Aristotle in order to have it represent an *ἀποδεικτικὴ ἐπιστήμη*. That is, there should exist a set of first true premisses[8] out of which all further truths may be inferred by (apodeictic) syllogisms. But syllogisms never count as propositions (*λόγοι ἀποφαντικοί*) nor do perfect syllogisms count as axioms (*ἀρχαί*), nor are the reductions of the syllogisms to perfect ones called 'proofs' by Aristotle.[9] Aristotle does not treat his syllogistic as a science. On the other hand, if the set-up of syllogistic would represent an art in the strong sense of a *διαλεκτικὴ τέχνη*, there should exist first premisses accepted for the sake of argument (*τόποι*), from which those propositions about which the argument is concerned follow by (dialectic) syllogisms. It is obvious that the apparent axiomatic treatment of Aristotle's syllogistic does not comply with these specifications, either. Syllogisms are used both for the sciences and for the arts, but they cannot themselves belong to either of them.[10] Consequently, a syllogism which is defined twice, in the

Prior Analytics and in the Topics, as a "linguistic expression (*λόγος*) in which, something having been posited, something other than the underlying results necessarily through the underlying"[11] should neither be read as a theorem nor as a rule, though both possibilities have been adopted alternatively through the centuries.[12]

This view gets further support by observing Aristotle's own argumentation on behalf of his choice for dealing with the objects of the Analytics. Instead of using the terminology of apodeictic or dialectic reasoning, which would throw some light on Aristotle's own opinion as to where to place the Analytics, he uses 'analytically' (*ἀναλυτικῶς*) instead of 'apodeictically', and 'logically' (*λογικῶς*) instead of 'dialectically' together with the interesting feature that most of his arguments on a certain point appear twice, once framed as an analytical one, and then as a logical one.[13] This indefiniteness on the status of the arguments for the set-up of argumentation itself should not really give rise to surprise. A far more detailed investigation is needed to free the beginnings of logic from the air of circularity. The reason syllogisms are treated neither as theorems nor as rules is simply that in a way they are indeed *both* theorems and rules, depending on the level of argumentation. They can, any one of them, be considered as rules of inference[14] – the syllogistic method in use is justly called a *συλλογιστικὴ τέχνη* by Aristotle[15] – but as soon as the syllogisms are not considered with respect to their producing something out of something, but as entities *sui generis*, those rules of inference may be transferred into (logical) implications, i.e. three-place (meta)propositions on propositions, and thus theorems.[16]

Hence, syllogistic in the sense of a theory of the valid rules of inference may be taken as an early anticipation of the position held by the proponents of effective logic, now in a refined version: effective logic is to be considered as a theory, i.e. a system of truths, about the universally admissible rules within arbitrary systems of rules of action.

Naturally, in the course of history, syllogistic has been treated the other way round, too. For example, according to the most influential diplomatic vote of scholasticism, which can be found in the *Summulae Logicales* (ca. 1250 A.D.) of Petrus Hispanus, who later became Pope John XXI, the definition of logic runs like the following: *dialectica* (i.e. logic) *est ars artium et scientia scientiarum ad omnium methodorum principia viam habens*;[17] and Duns Scotus gives an interpretation of this twofold determination: logic is a science '*respectu materiae ex qua constat*' and logic is an art '*respectu materiae in qua versatur*'. This distinction may now be taken as an anticipation of the position held by the proponents of classical logic, here again in a refined version, insofar as the system of tautologies can be enumerated by a calculus, i.e. a system of rules. More in the line of Aristotle, the more radical schoolmen such as Buridan in his '*Summa de Dialectica*' just dropped any mention of logic as a science and kept only its characterization as *ars artium* which, therefore, leads again to the position of the proponents of effective logic.

Hence, classical logic is the result of starting with arbitrary theories that obey the axiomatic method by concentrating on the forms of truths within arbitrary domains and then formalizing this system of formal truths by means of some calculus, thus getting a praxis on top of the theories. Effective logic, on the other hand, starts with arbitrary calculi built up by the constructive, i.e. genetic, method, and proceeds to a theory about the generally admissible rules within the calculi – a theory which can afterwards likewise be formalized; here we have a theory on top of the praxis.

Now it looks almost like a matter of taste how one is going to choose the level for a reasonable beginning of formal logic. Yet, the following constructions claim that there is an adequate solution of the conflict between a logic of rules and a logic of theorems and, more generally, a proper approach to the problem how praxis and theory interact in the case of logic.

II

The starting point is again very close to the actual origin of logic in antiquity. With Aristotle, and even more with Plato, logic – or rather dialectic, the term being a strong hint by itself – had to provide the means by which sound argumentation could be distinguished from unsound argumentation.[18] This has been a practical necessity in face of the highly developed sophistic technique to provide proofs for arbitrary theses on demand. And indeed, if it is granted that any scientific activity, be it on practical or on theoretical matters, is characterized as scientific by the fact that there is a justification available for each and every assertion put forth in the course of this activity (the possible linguistic articulations of non-linguistic acts included!), there is no other basis for the construction of logic than to look for a methodical *introduction* of the linguistic elements of assertions and from there to proceed to the *use* of assertions within argumentations.

Such an introduction of elementary linguistic elements shall be called *primary praxis* and will be executed within properly stylized teach-and-learn situations for these elements. As far as simple singular and simple general terms are concerned, the details of the procedure do not bear upon the set-up of formal logic. They have been discussed extensively elsewhere.[19] For our purposes, it is sufficient to remark that introducing words by means of teach-and-learn situations guarantees their public *understandability*. Furthermore, it should be clear that the determination of a primary praxis in the given sense is a process '*post hoc*', something man does in order to gain precise knowledge concerning his abilities and their limits after he has used speech and other acts meaningfully in the context of life.

The introductions in question do not each constitute a '*creatio ex nihilo*', they are rather '*re*creationes', that is a system of methodical reconstructions of that which has already been said and done. Another feature of the primary praxis is important: due to the teach-and-learn situations connected with the introduction of terms, there is no difference between the situation articulated by means of the terms, and the situations in which those terms are used. No use of terms other than introducing terms has as yet been the object of consideration. But this, of course, is a trivial part of human speech. The special power of linguistic communication becomes apparent only if the situations which underlie words, phrases, or sentences are different from the situations in which these words, phrases, or sentences are used. In that case, understandability of the linguistic expressions is not enough, a special link between the two situations is needed to secure the proper function of language. This link is provided by the detailed reconstruction – again through teach-and-learn-situations – of possible uses of linguistic expressions after their introduction. Any such introduction of a use of linguistic expressions different from the introduction itself shall belong to the *secondary praxis*, e.g. the use of terms as wishes, questions, or propositions. The way this is done guarantees the public *justifiability* of linguistic expressions in addition to their understandability.

The special act of asserting propositions (as distinguished from their use, e.g. in story-telling) involves a justifying procedure within the secondary praxis – a procedure which has to be introduced together with the use of terms as assertions – such that the validity of assertions can be defined by means of this procedure. It is even possible to distinguish words and sentences along these lines: if the situation articulated by a linguistic expression coincides with, or is at least part of the situation in which that expression is used, only its understandability is of concern, and the linguistic expression shall count as a word; but if those two situations are wholly different, both understandability and justifiability have to be secured and the linguistic expression shall count as a sentence.

This now is the exact point for characterizing the justifying procedure of assertions as a *dialogue*, or an argumentation between two partners. To assert a proposition makes sense only if there is someone on the other side, albeit fictitiously, who either denies or at least doubts the asserted proposition. But it is not enough merely to argue about propositions, there must exist precise stipulations on the rules of argumentation, rules which, in a way, define the exact meaning of the proposition in question.

A proposition shall be called 'dialogue-definite' under the condition that the possible dialogues on this proposition are finished after finitely many steps according to some previously stipulated and effectively applicable rules of argumentation, such that, at the end, it can be decided who has won and who has lost. Hence, dialogue-definiteness of propositions means that the relevant concept of a dialogue is decidable. And it is this concept of dialogue-definiteness that is to replace the age-old value-definiteness as the characterizing feature for linguistic expressions to be propositions.[20]

Further considerations will show that the class of value-definite propositions is indeed a proper subclass of the class of dialogue-definite propositions and that, therefore, the justifiability of propositions as introduced by the dialogue procedure does not coincide with their verifiability. Yet, 'truth' and 'falsehood' for propositions can now be defined on the basis of the *dialogue-game* associated with each proposition. Such a definition marks the beginning of a *theory* about (primary and secondary) praxis, insofar as (meta) propositions on the actions within the praxis get introduced. In a certain sense, even the secondary praxis itself contains a theoretical element, namely the propositions themselves, which get their meaning by the rules of argumentation about them. Hence, it might be appropriate, at least for the systematic purposes of the whole set-up, to distinguish an object-theory (the class of propositions introduced within the secondary praxis) from a metatheory (about primary and secondary praxis), the propositions of which cannot, of course, exist without the same pragmatic foundation as the propositions on the ground level. At this early stage, already, the interaction of praxis and theory is far more complicated than the usual presentation of logical theories permits us to suppose.

As a preparation for defining 'truth' and 'falsehood' for propositions, it is useful to observe that win and loss of a dialogue about a given proposition will in general depend upon an individual play of the game and will not be a function of the proposition alone. But the strategies of either player of the game are invariant against the choice of arguments of the other player. Hence, a proposition A shall be called '*true*', iff there is a *winning-strategy for* A; this means that the player who is asserting A – the *proponent* P – will be able to win a dialogue on A independently of the choice of arguments of the *opponent* O.

Accordingly, a proposition A shall be called 'false' iff there is a *winning-strategy against* A, i.e. the opponent can win a dialogue on A independently of the moves of the proponent.[21] I have shown elsewhere[22] that the dialogue on the metaproposition 'A is true' coincides with the dialogue on A itself, which means that the traditional condition of adequacy for any definition of truth is satisfied: 'A is true iff A'.[23]

The next step of the theory about the (primary and secondary) praxis is concerned with the justification of the rules of argumentation that constitute the secondary praxis. Again, this is done with a few accompanying remarks to the following proposal of a *structural rule* for dialogues, because space does not permit extensive elaboration on that point here.[24]

(D1) Dialogues about propositions consist of *arguments* which are put forth *alternatively* by an opponent O and a proponent P. The arguments follow certain *rules of argumentation* that belong to the game such that each play ends up with *win* or *loss* for either player.

(D2) With the exception of the improper initial argument, each argument either attacks prior ones of the partner or defends those of one's own upon such an attack, but does not act simultaneously in both ways: the proper arguments split into *attacks* and *defenses*.

(D3) Attacks *may* be put forth at any time during a play of the game (rights!).

(D4) Defenses *must* be put forth in the order of the corresponding attacks (upon which the defense answers), yet *may* be postponed as long as attacks can still be put forth: always that argument which has been attacked last without having been defended yet, has to be defended first (duties!).

(D5) Whoever cannot – or will not – put forth an argument any longer, has *lost* that play of game; the other one has *won* it.

(D1) is obviously not in need of further explanation; (D2) may be accepted as defining the special dialogue character of the game; and (D5) codifies equally current rules of win and loss. The only items in need of some further comments are (D3) and (D4), which regulate rank and order of attacks and defenses. With respect to the generality of rules, the right to attack shall not depend on a special position reached during a play of the game and, hence, shall not become void until the end of each play.

On the other hand, the given order of defenses is a consequence of the stipulation in (D1) to argue alternatively together with the rule of win and loss in (D5), if to both players is guaranteed that neither must defend upon an attack unless this attack has been defended first upon a counter-attack.

Now, (D1)–(D5) are not sufficient to secure finiteness of the individual plays of the game. There is lacking a regulation on the number of attacks permitted against a single argument during a given play. Yet, since any choice of bounds would be arbitrary, it might be accepted as reasonable that this choice should become part of the dialogue-game itself. After the initial argument has been laid down by *P*, first *O* shall choose a natural number *n* as the maximal number of attacks to be directed against a single argument of *P*, then *P* shall choose a natural number *m* analogously. Only now the proper dialogue about the initial argument may start obeying the following additional stipulation of the structural rule.

($D6_{n,m}$) During a play of the game, any argument may be attacked by the opponent at most n-times, by the proponent at most m-times.

Now, in order actually to play a dialogue-game according to the given rules, the rules of argumentation in (D1) have to be specified. This can be done by laying down a *schema of attacks and defenses*, which shows all possible attacks against an argument as well as all possible defenses of this argument upon each of these attacks. And, in general, this specification is possible only by special reference to the internal structure of the propositions concerned: the structural rule is purely formal in the sense that no special knowledge about the proposition is needed, whereas the rules of argumentation are material in so far as they have to make use of the actual set-up of the propositions, their 'content' in the terminology of traditional philosophy.

Yet, there is a possibility of determining special rules of argumentation that are, in a way, formal, too, namely, those that make use only of the fact that propositions may be composed out of subpropositions. This leads to the concept of *logical composition*, which in turn affects the introduction of further terms into the primary praxis, the so-called *logical particles*. A proposition A shall be called 'logically composed' out of propositions from a class K of given dialogue-definite propositions, if the schema of attacks and defenses associated with A contains only propositions from the class K. By means of such special rules of argumentation, the *particle-rules*, any dialogue about A is reduced to dialogues about the subpropositions of A: A is dialogue-definite, too.

Easy combinatorial considerations show[25] that for a complete survey it is sufficient to discuss unary, binary and infinitary logical particles only, under the condition to restrict the schemata in question to those that contain each subproposition just once and that use as further attacks certain non-assailable orders for defense or *doubts*, symbolized by '?' with added indices. Scheme 1 will be self-explanatory. (As notation for plays of the dialogue-game it has seemed useful to use two columns such that the rows are reserved for the consecutive attacks – from top to bottom with an index of the row number of that argument against which the attack is placed – together with an entry for the chosen defense – if any – upon that attack; to recover the order of moves one may enumerate the arguments, if necessary.)

As an example of a dialogue we will discuss the assertion $A_0 \leftrightharpoons a \rightarrow$

$b \overset{\cdot}{\to} b \overset{\cdot\cdot}{\to} b \to a \overset{\cdot}{\to} a$ for dialogue-definite propositions a and b. Furthermore, we will make use of one of the main results of the theory of dialogue-games, namely that the class of propositions for which there are winning strategies is not changed if O is limited to at most one attack against an argument of P, whereas P may choose any number of attacks against a single argument of O.[26]

	$*A$	attacks	defenses
position	$\ulcorner A$	?	A
negation (not)	$\neg A$	A	
	$A * B$	attacks	defenses
conjunction (and)	$A \wedge B$	1?	A
		2?	B
adjunction (or)	$A \vee B$	?	A
			B
subjunction (if-then)	$A \to B$	A	B
	$A \leftarrow B$	B	A
	$A \mathrel{-\!\!\prec} B$	A	
		?	B
abjunction (but not)	$A \mathrel{\succ\!\!-} B$	B	
		?	A
injunction (neither-nor)	$A \veebar B$	A	
		B	
	$*_x A(x)$	attacks	defenses
(all)	$\bigwedge_x A(x)$	$?n$	$A(n)$
(some)	$\bigvee_x A(x)$	?	$A(n)$
(no)	$\veebar_x A(x)$	$A(n)$	

Scheme 1

In this position, P must either have a winning-strategy for b in order to win by defending $a \to b$ with b upon the attack a or a winning-strategy against a. If, on the other hand, there exists a winning-strategy for a, P should have defended his second move upon the attack $b \to a$ instead of counter-attacking the first move of O with $a \to b$. Should O

have chosen the defense b upon the attack $a \rightarrow b$ as his fifth move, P would have attacked the third move with b, and any attempt of O to start now a sub-dialogue about this b of P would lead to an imitation of this sub-dialogue by P about the b of O (see Scheme 2).

O			P	
			A_0	
1) $a \rightarrow b \dot{\rightarrow} b$	(0)		$b \rightarrow a \dot{\rightarrow} a$	2)
3) $b \rightarrow a$	(1)			
		(1)	$a \rightarrow b$	4)
5) a	(3)			

Scheme 2

O must finally defend upon the attack b of P with a, and P in turn defends his second move with a. Any further attempt of O to try a sub-dialogue about this a of P results in an imitated sub-dialogue of P about the a of O.

Therefore, if a is value-definite, A_0 is true independently of the truth or falsehood of b. And it can be seen that there is a chance of winning-strategies which are *formal* in the sense that nothing need be known about the truth or falsehood of the prime propositions, as is the case e.g. for $A_1 \leftrightharpoons a \vee b \ddot{\rightarrow} b \rightarrow a \dot{\rightarrow} a$ (Scheme 3).

O			P	
			A_1	
1) $a \vee b$	(0)		$b \rightarrow a \dot{\rightarrow} a$	2)
3) $b \rightarrow a$	(1)		a	6)
5) a		(1)	?	4)

O			P	
			A_1	
1) $a \vee b$	(0)		$b \rightarrow a \dot{\rightarrow} a$	2)
3) $b \rightarrow a$	(1)		a	8)
5) b		(1)	?	4)
7) a		(2)	b	6)

Scheme 3

In these cases, the win of a play for P does not depend on the outcome of the dialogues about the prime propositions, the crucial point being only the possibility for P not to place a prime proposition as an argument until the same prime proposition has been placed as an argument by O.

A special rule for *formal playing* can, hence, be formulated:

($D7_m$) Prime propositions cannot be formally attacked; they may be put forth by the opponent without restrictions, whereas the proponent may only *take over* prime propositions from the opponent, each at most m-times, if m is the bound for the number of attacks against arguments of the opponent.

The restriction for P not to take over a prime proposition of O more than the maximal number of attacks he has chosen to place against a single argument of O during a play, is necessary in order to guarantee that the existence of a *formal winning-strategy* (i.e. a winning-strategy using formal playing only) for a proposition A is invariant against substitution of logically composed propositions for prime propositions within A.

On the basis of the construction up to this point, it is now possible to introduce the concept of logical (or formal) truth of propositions by the existence of formal winning-strategies for them. And a propositional schema is valid iff propositions bearing that schema are logically true.

This definition of validity for arbitrary dialogue-definite propositions works independently of any assumption on the truth or falsity of their prime propositions. It is, therefore, beyond the range of the classical theory of logical truth, even if the domain of propositions gets restricted both to value-definite ones and to quantifier-free ones. For, it is one thing to define the (classical) logical truth of a proposition A by the existence of (material) winning strategies for all propositions A^* which result out of A, if the set of prime propositions of A runs through all combinations of truth and falsehood with respect to these prime propositions, and it is another thing to define the logical truth of a proposition A by the existence of a formal winning-strategy for A. And, indeed, these two concepts do not coincide even within this restricted domain of propositions, as example A_0 already shows. The propositional schema $a \to b \mathrel{\dot{\to}} b \mathrel{\ddot{\to}} b \to a \mathrel{\dot{\to}} a$ is valid classically (with the special point that only the value-definiteness of a is needed), though there is no formal winning-strategy for it. The classical theory does not even permit the definition of this difference between general material truth and purely formal truth.

It is the pragmatic approach to formal logic by means of dialogue-games as has been sketched here which leads to the definability of a concept of formal truth for propositions which do not generally satisfy the classical condition of value-definiteness.

Hence, the dialogue concept makes it possible to define formal truth independently from material truth, whereas the classical theory is characterized by the reduction of formal truth to material truth, namely as general material truth.

At this late stage now it is reasonable to formalize the theory of dialogue-games and with this method to return to a praxis at a higher level, which as the praxis of the calculi of logic has been the usual starting point for contemporary logical theory. With the help of this formalization, it is possible to prove the main theorem of the theory of dialogue-games: The class of valid propositional schemata coincides with the class of intuitionistically valid propositional schemata.[27]

If furthermore all classically valid propositional schemata shall be gained by formal winning-strategies, this means fictitiously to assume the value-definiteness of suitable propositional sub-schemata of the propositional schema in question: any logically true proposition in the classical sense is logically true in the effective sense, if only suitable *tertium-non-datur* hypotheses are added. As an instructive example it is easily checked that there is a formal winning-strategy for the *classical disjunction* $\bigwedge_x a(x) \vee \bigvee_x \neg a(x)$ under the *tertium-non-datur* hypotheses $\bigwedge_x.\ a(x) \vee \neg a(x).$, $\bigwedge_x a(x) \vee \neg \bigwedge_x a(x)$, $\bigvee_x \neg a(x) \vee \neg \bigvee_x \neg a(x)$. Intuitionistic or effective logic is the logic of dialogue-definite propositions, two-valued or classical logic is the logic of the subclass of value-definite propositions. And it is quantification theory which shows the necessity for transition from the one to the other.

In any case, we can conclude that logic is primarily the theory of a structured praxis. The logic of antiquity has not in general been very conscious (at least in the eyes of its interpreters) of the pragmatic basis of logic taken as a science. As a substitute, one often uses the very misleading phrase of the 'ontological background' of ancient logic.

Yet, logic is, secondarily (by means of formalization), again a praxis of a structured theory. On the other hand, now, modern logic since Leibniz has minimized the importance of the possible theoretical basis of a calculus of logic, that is logic taken as an art. Presumably, one had doubts about the precision which could be imposed on a logic formulated only within ordinary language.

In both cases, there has been no clarity about the details of a step-by-step procedure from a praxis via a theory again to a praxis which, as we have tried to show, is the necessary minimum to get the means for an adequate solution of the current dispute on the true nature of logic.

NOTES

[1] L. E. J. Brouwer, 'Über die Bedeutung des Satzes vom ausgeschlossenen Dritten in der Mathematik, insbesondere in der Funktionentheorie', in *Z.f. reine und angewandte Mathematik* **154** (1925), 1–7, S. 2.

[2] Skolem's proposal of a strictly finite mathematics without any use of the quantors is an exception; its radical implications would deserve special discussion, cf. T. Skolem, 'Begründung der elementaren Arithmetik durch die rekurrierende Denkweise ohne Anwendung scheinbarer Veränderlichen mit unendlichem Ausdehnungsbereich', in *Skrifter utgit av Videnskapselskapet i Kristiania*, I. mathematisk-naturvidenskabelig klasse 1923, No. 6.

[3] A. Kolmogorov, 'Zur Deutung der intuitionistischen Logik', in *Math. Z.* **35** (1932), 58–65; P. Lorenzen, *Einführung in die operative Logik und Mathematik*, Berlin-Göttingen-Heidelberg 1955, [2]1969. Gödel has shown that an axiomatization of the concept 'beweisbar' (provable) within classical logic, somewhat different from Kolmogorov's nonformalized version, can be used as a representation of intuitionistic logic, cf. K. Gödel, 'Eine Interpretation des intuitionistischen Aussagenkalkuls', in *Ergebnisse eines Mathematischen Kolloquiums*, Heft 4 (1933), 39f.

[4] Cf. the review of the main results in S. C. Kleene, *Introduction to Metamathematics*, Princeton-Toronto-New York 1952, §81; especially important: J. C. C. McKinsey and A. Tarski, 'Some Theorems about the Sentential Calculi of Lewis and Heyting', in *J. Symb. Logic* **13** (1948), 1–15.

[5] A. Heyting, 'Die formalen Regein der intuitionistischen Logik', in *Sitzungsber. Preuss. Akad. Wiss., Physikal.-math. Klasse* (1930), 42–56, S. 45f.

[6] Cf. e.g. the paper 'Two Dogmas of Empiricism', in W. V. O. Quine (ed.), *From a Logical Point of View*, 9 Logico-Philosophical Essays, Cambridge, Mass., 1953, pp. 20–46.

[7] Cf. Brouwer, *op. cit.*

[8] The other conditions which Aristotle imposes on these first true premisses, the axioms (rather principles – *ἀρχαί*) as they are now called, are of no concern for our purposes, cf. An. post. 71b.

[9] This has been made a point in the convincing operative interpretation of Aristotle's syllogistic against the arguments of Łukasiewicz by K. Ebbinghaus: *Ein formales Modell der Syllogistik des Aristoteles*, Göttingen 1964.

[10] Cf. Arist. Met. 995a.

[11] Top. 100 a 25f, cf. An. pr. 24b 19f.

[12] For two modern proponents of either possibility cf. J. Łukasiewicz: *Aristotle's Syllogistic from the Standpoint of Modern Formal Logic*, Oxford [2]1957, and H. Scholz: *Abriß der Geschichte der Logik*, Freiburg-München [2]1959. Łukasiewicz interprets syllogisms as generalized subjunctions, Scholz reads them as rules of inference, e.g. PaQ, $QaR < PaR$ (*modus barbara*) becomes $\bigwedge_{PQR} . a(PQ) \wedge a(QR) \rightarrow a(PR)$, in Łukasiewicz, and PaQ; $QaR \Rightarrow PaR$ in Scholz.

[13] Cf. e.g. An. post 83b–84b, where there are two 'proofs' for the claim that to each science there must exist first true and undemonstrable principles.

[14] This is done successfully in the operative interpretation of Aristotle's syllogistic by Ebbinghaus as mentioned in Note 9; the complete syllogisms 'barbara' and 'celarent' are taken as intuitively valid rules; in addition, the interchange of terms in PeQ (*ἀντιστροφή*), the contradictoriness of PaQ and PoQ, and of PeQ and PiQ, and the contrariness of

PaQ and *PeQ*, all these formulated verbally by Aristotle, are given the form of rules; then all other valid syllogisms are 'provable' as admissible rules relative to the initial set of valid rules.

[15] Cf. Soph. Elench. 172a36.

[16] This is in accord with the characterization of arts and sciences in An. post. 100a 8ff; Arts are concerned with the world of coming-to-be and passing-away, sciences are concerned with the world of being.

[17] To the art-science dispute and its medieval background refers the discussion on modern operative logic (Brouwer, Wittgenstein, Kolmogorov, Lorenzen) by V. Richter, *Untersuchungen zur operativen Logik der Gegenwart*, Freiburg-München 1965.

[18] Cf. the first sentence of the Topics (100a 18ff), where the purpose of the treatise is characterized as "finding a method, by which we shall be able to argue (*συλλογίζεσθαι*) on any problem set before us starting from accepted premises (*ἔνδοξαι*) such that, when sustaining an argument (*λόγος*), we shall avoid saying anything self-contradictory." It was Kapp who showed convincingly the origin of Aristotle's syllogistic (still taken as a theory) in the actual sophistic discussions on public affairs, i.e. a praxis which was in need of regimentation; cf. E. Kapp, *Greek Foundations of Traditional Logic*, New York 1942.

[19] Cf. the second part of K. Lorenz: *Elemente der Sprachkritik. Eine Alternative zum Dogmatismus und Skeptizismus in der Analytischen Philosophie*, Frankfurt 1970; there may be found special references to competing proposals in W. V. O. Quine, *Word and Object*, Cambridge, Mass., 1960 and in P. F. Strawson, *Individuals. An Essay on Descriptive Metaphysics*, London 1959.

[20] The concept of a dialogue has originally been introduced by P. Lorenzen (cf. 'Logik und Agon', in *Atti del XII Congresso Internazionale di Filosofia* (Venice 1958), Firenze 1958f; 'Ein dialogisches Konstruktivitätskriterium', in *Infinitistic Methods, Proceedings of the Symposium on Found. of Mathematics* (Warschau 1959, Oxford 1961) for the purpose of a better foundation of operative logic; its further explication, especially with respect to a pragmatic foundation of the calculi of intuitionistic and two-valued logic is due to the author (cf. *Arithmetik und Logik als Spiele*, Kiel 1961; 'Dialogspiele als semantische Grundlage von Logikkalkülen', in *Arch. f. math. Logik u. Grundlagenforschung* **11** (1968), 32–55, 73–100).

[21] If, as usual, the validity of the logical principles is presupposed on the metatheoretic level, it would be possible from the validity of the saddle-point theorem for finitary two-person zero-sum-games to infer that propositions are *either* true *or* false. (Cf. C. Berge: *Théorie générale des jeux à n personnes*, Paris 1957); but without begging the question, there is only a practical meaning of 'either-or' on the metalevel available, i.e. decidability of choice, which cannot happen, since it is not generally decidable which part of the alternative holds; it is only decidable who has won a particular play of the game.

[22] Cf. K. Lorenz, *Dialogspiele als semantische Grundlage von Logikkalkülen*, p. 35f.

[23] Cf. A. Tarski: 'The Concept of Truth in Formalized Languages', in *Logic, Semantics, Metamathematics*. Papers from 1923 to 1938, Oxford 1956, 152–278, pp. 187f; for a discussion about the danger of semantic antinomies if this condition of adequacy is used as a schematic definition of truth, cf. K. Lorenz, *Elemente der Sprachkritik*, p. 44ff.

[24] For further details consult again the author's 'Dialogspiele als semantische Grundlage von Logikkalkülen', p. 37ff.

[25] Cf. *op. cit.*, p. 41ff.

[26] Cf. *op. cit.*, p. 85ff.

[27] Cf. for a proof the author's 'Dialogspiele als semantische Grundlage von Logikkalkülen'; another one in W. Stegmüller, 'Remarks on the Completeness of Logical Systems Relative to the Validity Concepts of P. Lorenzen und K. Lorenz', in *Notre Dame Journal of Formal Logic* **5** (1964).

JÜRGEN MITTELSTRASS

ON 'TRANSCENDENTAL'

1. DIFFICULTIES

"Logic is transcendental", Wittgenstein writes in the *Tractatus* (6.13) and, thereby, makes clear that his picture theory of knowledge also follows Kant's epistemological intentions. Wittgenstein calls logic 'transcendental' because it contains the condition for the possibility of representing facts: "The propositions of logic describe the scaffolding of the world, or rather they present it" (6.124). According to Wittgenstein, such a representation is a transcendental achievement, hence something that also, according to Kant, constitutes the essence of knowledge.

Wittgenstein's characterization of logic as transcendental is but one example of recent uses of the expression 'transcendental'. It is confusing that this concept is used in widely different systematic orientations and with reference to different contexts such as theory of argument, epistemology, and philosophy of science. What was for Kant the conceptual expression of an unmistakable new approach in philosophy, today partially determines the profile of philosophical schools, which otherwise have very little in common. Such circumstances require us to proceed cautiously and to exercise restraint – for example, by once again restricting the use of 'transcendental' to the systematic context of Kant's philosophy. However, I shall only follow my own advice to a certain extent. Not because, as yet another voice in the many-faceted philosophical canon, I am tempted to claim that I also have something new to say about 'transcendental'. Rather, it is my contention that, with respect to the use of 'transcendental', what matters is to save or further elaborate transferable insights. 'Transcendental' today has three chief meanings: defining (1) a figure of argument ('transcendental argument'); (2) certain procedures of the formation of knowledge ('transcendental method'); and (3) philosophical positions which can be characterized by using 'transcendental arguments' and/or 'transcendental methods' ('transcendental philosophy'). These meanings correspond in recent discussions to perspectives belonging to pre-

Robert E. Butts and James Robert Brown (eds.), Constructivism and Science, 77–102.

supposition-theory, metatheory or philosophy of science and epistemology. Such perspectives may be briefly characterized as follows.[1]

The concept of transcendental argument refers primarily not to achievements of constitution within the formation of knowledge (such as the use of 'transcendental' in Kant), but to the use of *tu-quoque*-arguments or to 'pragmatic self-refutation'.[2] By this, again, is not meant a logical-semantic relation of the following kind: negating p and asserting q results necessarily in a contradiction if p is logically presupposed by q. *Tu-quoque*-arguments or rather 'pragmatic self-refutations' involve a non-linguistic level: as a denoting term presupposes a thing, a *tu-quoque*-argument presupposes an act – you already *do* what you *deny in words*. It belongs to the 'transcendental' use of the argument of presupposition that acts, among them the utterance of a proposition, assume ('presuppose') the conditions for the possibility of their doing, that is: presuppositions are not propositions but acts. From this it follows, for instance, that if on the level of acts 'to be unconscious' and 'to speak about unconsciousness' exclude each other *pragmatically*, then on the level of propositions the statement 'I am unconscious' is necessarily false. This also makes clear that *tu-quoque*-arguments do not represent a *method* in a constructive sense, i.e. neither a method which generates theories, nor a method which generates principles. Only the *status* of an argument is characterized here as 'transcendental'. Therefore, an argument may be called 'transcendental' if it shows that something which is denied is (already) done, and it may further be called 'transcendental' if it exposes presuppositions (among them the presuppositions of the formation of knowledge) as *pragmatically fundamental*.

The original Kantian meaning of 'transcendental' has less to do with the idea of arguments in the sense of presupposition-theory than with the idea of method. It is true that Kant himself does not speak of 'transcendental method', but he makes clear that 'transcendental critique' represents "the method of knowledge through pure reason".[3] Correspondingly, the *Critique of Pure Reason* is called a 'treatise on method',[4] and "the determination of the formal conditions of a complete system of pure reason" a 'transcendental doctrine of method'.[5] For these reasons, Hintikka has recently argued against a too-narrow limitation of the meaning of 'transcendental' to *tu-quoque*-arguments, and has reminded us that the use of 'transcendental' refers to the formation of knowledge in a *constitutive* sense. By this is meant a method of *demonstrating* the possibility of a conceptual practice, particularly the possibility of the formation of 'theoretical' knowledge, and of *reconstructing* simultaneously the claim to validity of an actual practice, i.e. of actual knowledge. On the other hand, Hintikka's account also retains self-reference of *tu-quoque*-arguments as a property of transcendental methodological orientations: "The conclusion (the possibility of certain conceptual practices) is arrived at by reasoning which itself relies on these practices. The conclusion makes possible the very argument by means of which it is established."[6] Here, not only the status of arguments (as in the case of presupposition-theory) but the status of the formation of knowledge itself is at stake. Whenever one refers to this status through a system of propositions about propositions (in Kant's terms, about synthetic *a priori* propositions), corresponding analyses take on a *metatheoretical* character. Modern variants of a methodological approach take advantage of this state of affairs which, according to Stegmüller, also reflects the essential intentions of Kant's analyses, in the sense of a "metatheory of empirical knowledge".[7] Körner, on the other hand, undertakes a metatheoretical interpretation of Kant's analyses which tries to prove the impossibility of a transcendental foundation of the formation of knowledge.[8] According to Körner, the marking out of a certain categorical schema, to be found in Kant within the framework of 'transcendental deduction', cannot be justified. At the same time, the possibility of a 'transcendental method', at least in its Kantian form, is disputed.

Metatheoretical views, or, when applied to the formation of scientific knowledge, views subject to philosophy of science, are left behind by conceptions in which 'transcendental arguments' and 'transcendental method' serve to characterize certain philosophical positions (in the framework of epistemology). In this case, the paradigm is still Kant's own 'transcendental idealism'.[9] This idealism represents the *Lehrbegriff* (the doctrine) "that everything intuited in space or time, and therefore all objects of any experience possible to us, are nothing but appearances, that is, mere representations, which, in the manner in which they are represented, as extended beings, or as a series of alterations, have no independent existence outside our thoughts".[10] It is with such a *Lehrbegriff* that Kant refers to his 'Copernican revolution' according to which those *a priori* achievements of the subject of knowledge that belong to the capacities of (transindividual) subjectivity determine the objectivity of the knowledge of objects relating to experience. In contrast to *empirical* idealism, which concerns any representations (*Vorstellungen*) of empirical subjects, *transcendental* idealism deals, according to Kant, with representations which 'necessarily' result from the organization of our faculty of representation (*Vorstellungsvermögen*). Hence, the description of transcendental philosophy as a 'theory of *faculties of mind* (*Gemütskräfte*)'.[11]

Less far-reaching assertions and, therefore, a different '*Lehrbegriff*', specify new modes of transcendental philosophy, e.g., subsequent to Strawson, modes of a socalled analytical transcendental philosophy. In Strawson the reason for speaking about 'transcendental' is given by the analyses of concepts "the use and application of which are essential to empirical knowledge, and which are implicit in any coherent conception of experience which we can form".[12] That such a programme has little in common with Kant's idea of transcendental philosophy or 'transcendental idealism' is made clear by the fact that the achievements of the subject of knowledge, proved to be fundamental according to Kant's '*Lehrbegriff*', are completely disregarded in favour of categorical analyses and a theory about the identity conditions of empirical objects. I believe that, in a strict sense, there is today no non-Kantian transcendental philosophy, just as there is no non-Kantian transcendental method.

Furthermore, it is obvious that modern uses of 'transcendental' and the corresponding systematic orientations must concede some connection with Kant even if they are not explicitly linked to an interpretation of Kant. This means that it is still necessary to clarify Kant's position in any attempt at a systematic use of the term 'transcendental'. With this in mind, the following considerations serve to represent the conceptual context of 'transcendental'; to comment upon the problematic relationship between the concepts 'transcendental' and '*a priori*'; to offer a conception which allows us to work out or to complete Kant's theory in the direction of a further *a priori* ('*lebensweltliches Apriori*'); and finally to reformulate Kant's idea of 'transcendental critique' as defined by a programme of reconstruction.

2. CONDITIONS OF POSSIBILITY

One of the most frequently cited definitions of 'transcendental' in Kant runs as follows: "I entitle *transcendental* all knowledge which is occupied not so much with objects as with the mode of our knowledge of objects in so far as this mode of knowledge is to be possible *a priori*."[13]

In this definition all essential elements determining the use of 'transcendental' are assembled: its relationship to *a priori* knowledge; to the formula '*conditions for the possibility* of knowledge'; to the marking out of the *subject of knowledge*; and finally, at least indirectly, to aspects of *proof* and *foundation*. In spite of the ambiguity which, with regard to other parts of his work, burdens Kant's use of 'transcendental',[14] the following seems, from a terminological point of view, clear. According to Kant, 'transcendental' *inquiries* deal with knowledge, or rather a claim to validity of knowledge, formulated in 'synthetic *a priori* propositions', and with those *theories* which can be characterized by such inquiries ('transcendental idealism') or contain knowledge of the kind in question. Besides, and this was pointed out by Stegmüller and others, Kant anticipates theories which "produce *only* mathematical knowledge or theories which produce mathematical knowledge *in part* (natural sciences)",[15] whereby the term 'transcendental' once again acquires a 'metatheoretical' character. This characterization is not universal only because Kant at the same time in his epistemological analyses passes beyond analyses belonging to philosophy of science in a narrow sense; his 'architecture of pure reason' is not exclusively of the kind in which philosophy of science is interested. Therefore, 'transcendental' denotes "in essence, throughout the whole work, that is, through the first Critique, the reflection of reason on its own 'elements', its concepts and its *a priori* principles".[16]

Kant himself distinguishes with respect to his transcendental theory between a metatheoretical aspect and an epistemological aspect, without imparting, in his view, different meanings to the concept of the transcendental. This is indicated by the distinction between a 'synthetic-progressive' method which corresponds to the structure of the *Critique of Pure Reason*, and an 'analytic-regressive' method which corresponds to the structure of the *Prolegomena*.[17] Whereas the 'analytic-regressive' method *assumes* the actual validity of a theory (here Newtonian physics) and *reconstructs* its structure, the 'synthetic-progressive' method attempts to do without such an assumption and, therefore, does not orient itself explicitly with respect to a theoretical structure already achieved. Its ability of reconstruction is assumed 'universally', i.e. adequately for all theoretical formations which, in that case, play the role of examples. What at first sight appears to be an essential distinction by which Kant's philosophical theory divides itself into a 'transcendental' part and a 'metatheoretical' part, turns out to be in Kant's own view only a distinction between methodological perspectives. Indeed, the synthetic-progressive method remains dependent at least upon knowledge without whose recognition speaking about possible experience would not make any sense. Thus, the method itself contains a regressive element; and, as far as the analytic-regressive method in its orientation towards a theoretical *factum* not only represents this *factum* but supplements it by revealing its 'epistemological' pre-conditions, it contains a progressive element. If, as this interpretation suggests, 'transcendental', in the framework of Kant's terminology, denotes the reflection of reason upon its own elements, then this reflection takes place, within a synthetic-progressive method, *universally* upon 'concepts and *a priori* principles', and, within an analytic-regressive method, specifically upon 'concepts and *a priori* principles' of a *given theory*.

It is an additional constitutive element of 'transcendental' that 'concepts and *a priori* principles' as well as so called 'pure intuitions' (space and time) function as *conditions for the possibility* of experience. This element is often difficult to understand. Particularly, it is unclear how the formula 'conditions of possibility' relates to the *logical* distinction between necessary and sufficient conditions which itself is drawn from the analysis of if-then sentences (e.g. implications like $a \supset b$, $(\forall x)(a(x) \supset b(x))$, which are universal hypothetical sentences). The traditional reading which holds that the Kantian formula also focuses on logical relations, is problematic at least with regard to the additional distinction between an interpretation of if-then sentences according to a *logic of propositions* or rather *quantifiers* and to *modal logic* (an event E_1 is called a sufficient condition of another event E_2 if $\Box (E_1 \supset E_2)$ is valid, and it is called a necessary condition of E_1, if $\Box (\neg E_1 \supset \neg E_2)$ is valid). It is also problematic with regard to the interpretation of modal relations as *causal* relations between events.[18] In general, in Kant the realization of theoretical and pragmatic intentions is dependent upon the observance of certain rules or rather (again) upon such intentions which are already realized (as in the case of presupposition-theory). This would also be in accordance with the *constitutive* sense of the formula in Kant which cannot be confined to logical relations only. On the other hand, this sense confines conditions to those that are *conceptual* or *pragmatic* in kind; it thus excludes, for instance, taking into consideration historical, social or 'natural' conditions (such as the existence of the central nervous system).

In the framework of the *constitutive* sense of the formula (i.e. the sense given by a theory of constitutive relations), there occur aspects of 'transcendental' proof (of certain propositions or complexes of propositions) as well as aspects of 'transcendental' *foundation*, thereby problems of the *beginning*. These aspects, i.e., the specification of steps methodically leading to knowledge (proof) and the specification of a basis allowing methodical constructions (foundation), are not always strictly separated from each other. For instance, 'transcendental' proofs like the proof of the autonomist thesis in the discussion of the antinomy of freedom include references to 'foundational' facts (here the "power of beginning a state *spontaneously*"[19]); 'transcendental' foundations, in turn, include references to sequences of proof (e.g. in the manner of rule-governed steps in the case of the transcendental schemata[20]). In general, in Kant the conception of the transcendental seems to lead back from steps of proof to steps of foundation, i.e. exactly to pointing out 'conditions of possibility'.

Kant, indeed, is not content with this reduction. The synthetic *a priori*, the 'possibility' of which is systematically at stake here, is seen chiefly in basic structures of the organization of consciousness or the human apparatus of perception. In particular, it is the distinction between 'empirical consciousness' and 'transcendental consciousness' ('*Bewußtsein überhaupt*') that is supposed to reduce 'conditions of possibility', with a founding intention, if not to an empirical then to an 'original' *factum*, i.e. the "original unity of apperception"[21]: "The given intuition must be subsumed under a concept which determines the form of judging in general relatively to the intuition, connects empirical consciousness of intuition in consciousness in general, and thereby procures universal validity for empirical judgements."[22] Efforts have been made to use this background from the theory of consciousness in order to establish a Kantian determination of the meaning of 'transcendental'.[23] Whether this is also of systematic advantage must certainly be taken as highly problematic; systematically conclusive reconstructions of the Kantian concept of the 'original apperception' are still missing. Thus, it may be safe to assume that, from a systematic point of view, efforts towards reconstructing Kant's analyses of a synthetic *a priori* without using means that belong to a 'theory of consciousness' are much more convincing. Only such efforts, I believe, can put us into the position to join modern systematic insights *and* Kantian insights.

3. SYNTHETIC A PRIORI

If 'transcendental' denotes reflections of reason on concepts, (pure) intuitions and *a priori* principles, in regard to their role as 'conditions for the possibility' of experience, then a clarification of 'transcendental' essentially depends on a clarification of '*a priori*'. This is true not only with a view to Kant's definition of 'transcendental', but also, so far as *non-empirical* elements are here rendered prominent, with respect to any systematic definition of that concept.

It has already been indicated that, according to Kant's epistemological 'Copernican revolution', *a priori* achievements of the subject of knowledge prove to be responsible for generating the objectivity of the knowledge of objects relating to experience. Kant expresses this as follows: "that we can know *a priori* of things only what we ourselves put into them."[24] For that reason, there are particular propositions, namely synthetic *a priori* propositions, in which conditions (concepts, pure intuitions, principles) are formulated which constitute 'experience', i.e. the 'knowing of objects'. Still using Kant's terminology: knowing is 'objective' in that its objects are generated by us, and are therefore 'subjective', i.e. generated by the faculty of knowledge or 'consciousness'. Objects of knowledge are not 'things in themselves' but *representations* (*Vorstellungen*). They are *our* representations, and that is why the 'knowing of objects' amounts first of all to the understanding of our own 'constructions'. The object of knowledge (the sensible given) is structured by the (pure) forms of intuition (space and time) and (mediated by the schematism) by the concepts of the understanding (categories). This means precisely: only the intuitive and conceptual achievements of differentiation as well as the 'synthetic' achievements of the subject (of knowledge) constitute the object (of knowledge) and, in this way, referred to the particular achievement of knowledge, render 'objective' knowledge possible. Kant's first principle of 'transcendental deduction' gives expression to this: "The conditions of the *possibility of experience* in general are likewise conditions of the *possibility of the objects of experience*, and [. . .] for this reason they have objective validity in a synthetic *a priori* judgement."[25]

So much for Kant's account of the role of the synthetic *a priori* organizing experience. How is it possible, though, to achieve methodological access to a way of speaking about '*a priori*'-elements which also are to function as elements of the concept of the transcendental? Kant himself gains this access by an analysis of the two pairs of contrasting concepts: analytic – synthetic and *a priori* – *a posteriori* (i.e. *a priori* – empirical), which represent two distinctions that are completely independent of one another (except for the connection analytical – empirical).[26] The point is to determine the meaning of the connections analytic – *a priori*, synthetic – *a priori*, and synthetic – *a posteriori* (the connection analytic – *a posteriori* is excluded because of the definition of 'analytic' as an inclusive relation of concepts). In this context, Kant calls propositions '*a priori*' if they provide knowledge "independent of experience and even of all impressions of the senses"; and he calls them 'empirical' if they have their "sources *a posteriori*, that is, in experience".[27] Analytic propositions turn out on this analysis to be *a priori*, non-*a priori* propositions to be necessarily synthetic, namely non-analytic ("judgements of experience, as such, are one and all synthetic"[28]). The question remaining is to determine the meaning of the connection 'synthetic – *a priori*', i.e. to determine the possibility of synthetic *a priori* propositions. Such propositions are (1) synthetic, i.e. not fully determinable by analytic, conceptual explication, but they have (2) an *a priori* validity, i.e. they are not, despite their synthetic character, refutable in terms of the conclusion which follows from them (in Kant's own formulation: they are not refutable by experience). The answer to this question is offered by an arithmetical example, which shows that we are dealing here with propositions whose validity is not only based upon logical and definitional (analytical) factors. It is based rather upon *constructions*, which themselves are not based entirely upon logical and definitional determinations of the utilized concepts. Besides geometry and arithmetic there are, above all, definite foundational principles of physics (for instance, the so-called causal principle) whose validity is claimed to be derived from their synthetic *a priori* character.

It belongs to the essential results of Kant's analysis that the distinction between analytic and empirical, viewed from a methodological perspective, turns out to be incomplete. Moreover, Kant establishes the identity of '*a posteriori*' and 'empirical' (an essential contribution as viewed from the perspective of the history of concepts). By 'empirical' Kant means an *empirical-physical* proposition, i.e. a proposition which is based upon results of measurements. In this sense Kant also speaks of '*empirische Erkenntnisse*'. The distinction between empirical validity and *a priori* validity of propositions must therefore be understood in such a way that the empirical validity rests upon the results of measurement (especially measurements of length, time and mass), whereas *a priori* validity arises when the truth of any proposition does not rest upon such measurements. Regarding the concept of the synthetic *a priori*, established in this context, it is, of course, important that a methodological connection between empirical validity and *a priori* is constructed by Kant utilizing the claim that '*empirische Erkenntnisse*' (empirical knowledge) cannot be properly established independently of *a priori* knowledge. In contrast to its former use as a meta-predicate applied to *proof-theoretical* connections, the concept of the *a priori* acquires here a *foundational* aspect. The assertion is made that the empirical sciences such as physics in both their methods of measurement and their leading orientations (e.g. basic principles such as the causal principle) are dependent upon non-empirical propositions, whose validity is entirely *a priori*. This clearly emerges from the manner of speaking about the conditions for the possibility of experience as well as from the fact that '*a priori*' usually is intended by Kant to refer to the '*a priori*' part of empirical-physical knowledge (e.g. what he means by 'experience').

In following Kant's analysis without thereby committing ourselves to defending his position in detail, the following more precise definitions can be suggested: (1) *A proposition a is true a priori (or valid a priori), if its proof can be given without reference to the consequences which follow from a.* In this definition, which rests upon a synonymy between 'true' and 'proven', '*a priori*' is specified as a proof-theoretical meta-predicate, i.e. as a proof-theoretical *a priori*, insofar as it makes no reference to the kinds of consequences which follow from *a*. (2) *A proposition a_1 is true a priori (or valid a priori) only in relation to another proposition a_2 which results from certain definite measurements (in other words, a_2 is an empirical-physical proposition)*. This definition represents the foundational restriction introduced by Kant, which has been responsible for most of the subsequent epistemological controversies about the status of *a priori* proofs. Logically viewed, in both definitions the '*a priori*' is a dyadic relation between propositions; in definition (2) one proposition (or the propositional system a_2) is characterized as empirical in the above-mentioned sense.[29]

Thus, it becomes clear again that all *analytic* propositions constitute a part of *a priori* knowledge, whether they are *formal-analytic* propositions founded upon logic-definitional conventions, or *material-analytic* propositions founded upon certain terminological rules in addition to logico-definitional conventions. At this point Kant, with regard to the concept of the transcendental, puts forward an essential restriction: not all *a priori* knowledge can be characterized as 'transcendental'. 'Transcendental' knowledge is only knowledge founded upon synthetic *a priori* propositions – hence, according to Kant, not only rational physics but also geometry and arithmetic, as far as both deal with constructions in pure intuition. By way of contrast, analytic knowledge, although it is *a priori* knowledge, has no transcendental status: "Not every kind of knowledge *a priori* should be called transcendental, but that only by which we know that – and how – certain representations (intuitions or concepts) can be employed or are possible purely *a priori*. The term 'transcendental', that is to say, signifies such knowledge as concerns the *a priori* possibility of knowledge, or its *a priori* employment."[30] Therefore, only the foundational *a priori* establishes a *transcendental a priori* in the strict sense. And this makes good sense. For, what would be the reason, for instance, to suspect transcendental truth or a transcendental *a priori* lying behind the (formal-analytically true) proposition: '$x, y \, \varepsilon$ married $\supset x \, \varepsilon$' 'bachelor' (the term 'bachelor' defined by: $x \, \varepsilon$ bachelor $\equiv_{df} x \, \varepsilon$ male & $\neg \, (\exists y)$ $(x, y \, \varepsilon$ married$)$?

4. LEBENSWELTLICHES APRIORI

By characterizing the synthetic *a priori* as transcendental *a priori*, we have articulated distinctly the constitutive meaning of Kant's concept, as against different forms of *a priori* knowledge (e.g. the analytic forms). This concept is also clear from a *logical* point of view, in that, within the transition from a proof-theoretical meaning of '*a priori*' to a foundational meaning, the distinction between analytic and empirical is shown to be incomplete. Kant's supplementation of that distinction (whose incompleteness has been recently discussed anew in light of the problem of theoretical and dispositional concepts) by the concept of the synthetic *a priori* makes allowance as well for the particular circumstances of mathematics and physics within whose limits *synthetic* determinations constitute the objects of knowledge. In a further analysis within philosophy of science the synthetic *a priori* of Kant in this sense leads to the distinction between *formal-synthetic* propositions (i.e. propositions whose truth can be established by referring to logico-definitional conventions and pragmatically justified symbolic constructions) and *material-synthetic* propositions (which additionally provide for ideal norms of action, i.e. 'ideative rules', e.g. in the case of the Euclidean plane). In the framework of such a conceptual elaboration of the synthetic *a priori*, formal-synthetic propositions characterize arithmetic, material-synthetic propositions geometry and protophysics built up by means of geometry (being a theory of the quantities of measurement or of the measuring instruments).[31]

In view of these distinctions Kant's attempt to embody the transcendental *a priori*, particularly in its transcendental use, in basic structures of the organization of consciousness or the human apparatus of perception is, again, systematically unclear. Here, the epistemological framework, in which Kant's considerations take place and which may be characterized as a philosophy of (transindividual) subjectivity, turns out to be too restricted; Kant's efforts to remain within that framework inevitably lead from analyses belonging to a theory of constitution into material theories (about consciousness or perception). To put it differently: Kant gets drawn into such theories because (1) he did not yet consider the possibility of a linguistic reconstruction of constitutive structures and because (2) he no longer saw the 'Aristotelian' way of embedding scientific concepts of experience in an analysis of pre-theoretical circumstances.

It would, of course, be wrong to assume that Kant's concept of experience is only committed to empirical-physical knowledge and is, therefore, exclusively a meta-theoretical concept. This is made clear, among other things, by the analyses of the 'transcendental aesthetic' in the *Critique of Pure Reason*. For, if Kant, within these analyses, also calls intuitive space a 'condition for the possibility' of experience, then what is meant by experience includes both pre-theoretical (therefore, up to a point, also 'Aristotelian') faculties of orientation and empirical propositions of experimental physics (in Kant's own words: '*empirische Erkenntnisse*'). This may be briefly explained.

With respect to the 'transcendental aesthetic', the issue is how to prove that the faculty of spatial orientation does not refer to empirical knowledge gained in a way similar to knowledge related to or founded upon objects of perception. The same is true for extensive bodies and spatial forms as well as, by analogy, faculties of temporal orientation and their realizations (e.g. steps of action like succession and repetition). This proof is achieved by explaining not only that space cannot be perceived and is, therefore, not an empirical object,[32] but also that space always, as a representation, accompanies every perception and, furthermore, every experience, because perceiving an object presupposes that representation. According to Kant, space, just as time, being a 'pure form of intuition' comes before "all appearances and before all data of experience". It is, indeed, what makes "the latter at all possible".[33] This marks the discovery of a *non-conceptual a priori* organizing the faculty of orientation and, in addition, the order of 'appearances'; this non-conceptual *a priori* is then identified by Kant as part of the methodological foundations both of physics and mathematics. Kant regards space as 'pure' intuition, intuition *a priori*, because concepts are the meaning of predicates on any number of 'examples', whereas one speaks of space only with reference to one and the same intuitive space, and that without referring to empirical intutitions. The observation that space as pure intuition belongs to the foundations both of arithmetic and geometry is to be understood in the following way: the propositions in arithmetic and geometry are founded upon intuitive constructions, i.e. upon the construction of spatial forms and sequences of intuitive figures. Thereby, however, a theory of intuitive space like the theory which Kant, for the first time, tries to establish independently, from a methodological point of view, of empirical-physical and formalist-mathematical theories, provides both a more specific determination of the concept of *construction* (as construction in intuitive space) and the proof that scientific and pre-scientific (or pre-theoretical) experience is dependent on the conditions of their spatial production or spatial occurrence.[34]

If one disregards the remaining vagueness about the different status of the *intuitive structure* of scientific as opposed to pre-scientific (pre-theoretical) experience as well as the connection with a theory of perception which is problematic from a systematic point of view, one can say that Kant's analysis of the 'pure forms of intuition' space and time provides, within the explication of a non-conceptual *a priori*, a first approach towards the definition of a *lebensweltliches Apriori* of scientific theories. As far as 'space' is concerned, this *a priori* takes effect both in the faculty of spatial orientation and in the practice of constructing spatial forms; it is, therefore, also appropriate to found the synthetic *a priori* not upon a theory of consciousness but upon a reflection on the *pragmatic* achievements of orientation and construction. To speak more accurately, the synthetic *a priori* does not itself have a pure 'theoretical' structure; rather, it is composed of pre-theoretical ('*lebensweltliche*') and scientific elements organizing 'empirical knowledge'. Passing systematically beyond Kant's analysis, this can be proven by reconstructing a *graded a priori* of empirical theories. The result of such a reconstruction is[35] that on the basis of a *pre-theoretical a priori* which, in an 'Aristotelian' sense, can be characterized as being a 'nest' of pragmatic habits of orientation and action, is built up a *measuring-a priori* which itself sets the standards for the procedures of measurement which guarantee the objectivity of a measuring (empirical) theory and thus form the basis of hypothetico-deductive empirical theories. In this reconstruction the synthetic *a priori* of empirical theories proves to be much more complex than Kant's analysis brings to light.

To illustrate what is meant by a pre-theoretical or '*lebensweltliches*' *a priori*, let me briefly point out the 'epistemological' role of *predication.*[36] A kind of elementary knowledge and activity, on which also the methodological construction of theoretical knowledge depends, has to do, in the first place, with the learning of, and the control over, *distinctions* and *orientations*. By means of actions of distinguishing and actions of orientating there is constituted epistemologically a pre-theoretical practice. It can be characterized as an integrated teaching- and learning-situation. In such a situation simple speech-acts (such as making distinctions) and non-linguistic activities (such as visualizing objects in space) are *pragmatically* controlled. The subordination of such practical activities to *norms* (e.g. the subordination of a factual practice of using distinctions, however complex it might be, to an introductory practice) leads in each case to a reconstruction of the distinctions-knowledge given with this practice. This reconstruction already includes the reconstruction of elementary proofs, such as the 'argumentative' establishment of elementary propositions, by means of which the first simple questions of validity can already be answered. Conversely – and this is the essential point – every theoretical proposition depends upon this elementary practice of distinguishing, of using distinctions, and of arguing for and against claims dependent upon distinctions. There is no 'beginning' in the strict sense for theoretical knowledge, which could stand independently of distinctions or which could dispense with a normative formulation of elementary speech-acts. Indeed, the concepts of theoretical discourse, which Kant still tried to state 'directly', i.e., without a preceding clarification of the practice of predication itself, are founded upon such an (elementary) practice of predication.

To formulate the point methodologically, theoretical knowledge is based upon the knowledge given by elementary predication, i.e., by elementary propositions and their logical nexus, corresponding to elementary predication. In addition to the genetic primacy of a knowledge of distinctions and orientations available within the *Lebenswelt*, there is the logical primacy of elementary predication with respect to complex propositions which build up theories and to specialized methods of proof which are relevant to them. Thus, elementary predication, epistemologically conceived, is an *a priori* part of all knowledge, since (1) the theoretical knowledge does not include either explanation or proof without making use of elementary predication. (2) No *a priori* proof of a theory can be given, in which we do not already find, as *a priori* elements, certain pre-theoretical activities of distinction and orientation. I call this *a priori*, which is part of the *lebensweltliches Apriori*, the *distinction-a priori* (*Unterscheidungsapriori*).[37]

Using Kant's terminology, a 'transcendental' status of predication has been reconstructed here: predication is a linguistically fundamental act; in the strict sense, one cannot go beyond it. Every reflection on predication, i.e., every philosophical or linguistic theory of predication, already uses predication. In other words, (1) predication is, in fact, a *beginning*, it cannot be 'founded' upon something else. (2) Predication yields the 'condition for the possibility' of common orientations including theories, since such orientations operate by making distinctions. Where there are no distinctions made, nothing is said. On this fact also rests, essentially, Wittgenstein's thesis on the transcendental character of 'logic' quoted at the beginning. It has been shown that the picture theory in the *Tractatus*, where this thesis comes from, can be interpreted as being a theory of predication.[38] As a theory of predication, in the sense of what has just been said, Wittgenstein's picture theory would comprise the 'condition for the possibility' of representing facts. Kant would then be turned upside down (or right side up) – from the head (the concept of subjectivity based upon a theory of consciousness) to the feet (the concept of a *lebensweltliches Apriori* based, among other things, upon a theory of predication).

5. PRAGMATIC RECONSTRUCTION

There are to be found in Kant some explications of the concept of the synthetic *a priori* which obviously do not require such a procedure. Whereas the conception of a *lebensweltliches Apriori*, in the sense of a theory of predication, can only most indirectly be related to Kant, things are entirely different in the case of the *intuitive structure of mathematical constructions* analyzed by Kant. Here, Kant himself leaves the level of propositions and conceptual explication; he instead argues on a *pragmatic* level, as was indicated earlier in our discussion of Kant's theory of intuitive space. It is not necessarily an argument against its systematically elementary character that the exposition under the title *The schematism of the pure concepts of understanding*[39] belongs to the most difficult part of the *Critique of Pure Reason*.

Kant's explication of a non-conceptual *a priori*, as we have already mentioned, starts from the assumption that 'mathematical knowledge' must be based upon 'a pure intuition' "in which all its concepts can be exhibited or constructed, *in concreto* and yet *a priori*".[40] 'Transcendental schemata', in this sense, represent *rules* or *methods of construction*.[41] Accordingly, in Kant's explanation, '*number*' is the schema of the category 'quantity', characterized by two rules of construction

$$\Rightarrow |$$
$$n \Rightarrow n|$$

In this arithmetical calculus the first rule signifies an opening rule (a rule without premises) by which the basic figure '|' is constructed. The second rule signifies the transition to another figure: if n, i.e. the premise of the rule, is constructed, then it is permitted to construct $n|$ too, i.e. the conclusion of the rule. The rule-arrow '$\Rightarrow$' signifies here a *practical* 'if-then'. Whenever an act has been performed resulting in a figure n, another act, by which '$n|$' is constructed, shall be permissible according to the rules.

Kant's explanations make clear that this meets his systematic intentions. 'Transcendental schemata' are supposed to establish the relationship between 'pure concepts of the understanding' (*reinen Verstandesbegriffen*) and 'appearances' (*Erscheinungen*), in such a way that the problem of *applying* conceptual means to what is given in sensible perception can be solved. It is "just as necessary to make our concepts sensible, that is, to add the object to them in intuition, as to make our intuitions intelligible, that is, to bring them under concepts. These two powers or capacities cannot exchange their functions. The understanding can intuit nothing, the senses can think nothing. Only through their union can knowledge arise."[42] This connection is achieved by the 'transcendental schemata', since these schemata are both 'intellectual' and 'sensible'.[43] They are 'sensible' in that they refer to the structural properties of sequences of acts like beginning, succession and repetition, but not, however, 'pictorially' to certain intuitive acts: a 'transcendental schema', that is, "the schema of a *pure* concept of understanding can never be brought into any image whatsoever. It is simply the pure synthesis, determined by a rule of that unity, in accordance with concepts, to which the category gives expression."[44] Accordingly, for Kant, number is "a representation which comprises the successive addition of homogeneous units".[45] 'Schematic' here signifies the *repetition* of an act starting with a *beginning*, from which point the pertinent method of construction can make use of different forms of realization 'within sensible intuition'. In the case of the arithmetical calculus, for instance, of sequences of lines or other figures (the identity of figures which here represent ciphers for numbers is guaranteed by two other rules $\Rightarrow| = |$ and $m = n \Rightarrow m| = n|$). It is important here, according to Kant, that to the '*concept*' of number corresponds not a *picture* but a *method of construction* 'within intuition'. And this schematic method of representation is 'transcendental', because with this method the generic conditions of concepts 'within intuition' are given.[46] The propositions gained by such a method were classified earlier as *formal-synthetic* propositions, as opposed to the material-synthetic propositions of geometry.

Thus, when speaking about 'transcendental schemata' Kant is concerned in general with the relationship between 'concept' and 'intuition', and he is concerned in particular with the reconstruction of *pragmatic foundations* ('intuitive structures') in mathematics. Those pragmatic foundations are included in structures of action. If we apply this to what has been said about the *distinction-a priori* the result is: Whereas learning a schema of action (in the case of arithmetic, for instance, joining together and iteration) does not presuppose the validity of any 'theoretical' proposition, it makes it possible, as in the case of arithmetic through norms concerning the construction of figures (ciphers) according to rules, to generate 'theoretical' objects, in this case the natural numbers. With respect to their foundation, then, the *construc-*

tions of mathematics turn out to be *reconstructions* of a pre-theoretical ('pragmatic') practice. But since such reconstructions belong to the concept of construction, there is in Kant a presupposed *transcendental pragmatics*.

In this pragmatics Kant's analyses reach a non-conceptual and pre-theoretical *a priori*, also with regard to particular knowledge (as demonstrated in the light of the arithmetical calculus), without explicitly identifying this *a priori* within the concept of the synthetic *a priori*. At the same time, in this context, the concept of the transcendental acquires certain aspects which allow us to transform it into new considerations. What is meant here is not the usual characterization of 'transcendental' or '*tu-quoque*-arguments' as mentioned at the beginning (arguments showing that what is denied is already done, or exposing presuppositions as incapable of being gone beyond pragmatically). Among other things, such arguments do not solve the problem of the beginning. Unlike the idea of the transcendental in Kant, the schematism of number is not only 'transcendental' in the sense of its character in constituting knowledge. It is, rather, a beginning as well, in the manner of rules of construction. I use the concept of reconstruction in such a way as to include in its meaning both *logical* (or *rational*) and *pragmatic* elements.

Today one speaks about 'reconstruction' mainly with regard to the analysis of theories. In a wider sense this also means with regard to the analysis of the genesis of theories and of scientific developments. Here, 'reconstruction' signifies a procedure which operates on the level of propositions and conceptual explication. Hence, it is characterized as 'rational' or 'logical'. The paradigm of rational reconstruction in this sense is Carnap's programme for a revision of unclear scientific languages by exact linguistic constructions.[47] This programme was carried on by Reichenbach, who also supplemented it with normative elements under the concept of 'logical analysis' or 'analysis of science'.[48] Today it is particularly represented by Stegmüller, who again, quite in Carnap's sense, takes it as the aim of 'reconstructive' explications to transform conceptual determinations into formal structures.[49] In general, one may say that a reconstruction can be considered as successful or adequate, if a construction K', substituted for a given conceptual context K, not only reproduces K correctly in all its essential parts but also, at the same time, fulfils the intentions pursued by K better (at least not worse) than K.

This frame of reference no longer suffices, however, if the concept of reconstruction is supplemented by *pragmatic* grounding-relations, i.e. if one departs from the level of propositions and conceptual explanation in this respect. This is exactly what happens not only in Kant with his conception of schematism, but also in recent constructive theories of mathematics and physics.[50] At the same time, the concept of *pragmatic reconstruction*[51] leads to the idea of a justified construction of scientific theories in place of the idea of an inevitable beginning with a theoretical framework (a position represented by Popper). This framework, so it is claimed, may be left behind in favour of another framework but it is not capable of being justified itself via reconstruction. This is, in a certain sense, impossible if, at the same time, the convention is effective, that reconstructions must not leave the level of propositions and concepts. If this convention is not effective we are again, within the modern concept of reconstruction, up against Kant's problem of a synthetic *a priori*.

Does that mean that the term 'transcendental' should find a new use in philosophy of science? On first sight this seems to be a reasonable assumption. Within pragmatic reconstructions, i.e. within reconstructions in which actions subject to norms or schemes of action determine beginnings and also, to some extent, the further construction of theoretical structures, there exist pragmatic conditions, which cannot themselves be based 'upon theory'. As an example one may think of a definition of geometrical concepts according to methods of construction for spatial forms.[52] In Kantian terms this could also be expressed by saying that we are dealing here with 'transcendental' conditions; correspondingly it has been said that there is, in fact, in Kant already a transcendental pragmatics. These conditions are, furthermore, part of a graded *a priori*, not part of a contingent *factum*. This is expressed by the fact that within the context of that construction whose pragmatic part they are, they cannot be replaced by something else (particularly not by theoretical structures). The other aspect of Kant's use of 'transcendental' is also taken into account, i.e., noticing such conditions *forces* a certain construction.[53]

There are, however, good reasons not to extend use of the term 'transcendental' to rational (or logical) and pragmatic reconstructions. To these reasons belong: (1) The requirement to demarcate, always anew, the use of 'transcendental' with respect to those intentions and systematic elaborations of Kant which shall not be taken over. To those might belong, for instance, the whole epistemological framework appertaining to a philosophy of subjectivity, at least those aspects whose elaboration belongs within the boundaries of a philosophy of consciousness. After all, Kant's own concept of schematism shows that it is unnecessary, with respect to the achievements of actions subject to norms or schemata of action, to include the 'unity of consciousness' as part of the 'conditions for the possibility' of arithmetic. (2) The requirement to satisfy the impression that there is something like a 'transcendental method'. There are methods whose status in view of certain results, as in the case of intuitive constructions within mathematics, might be called 'transcendental', but it is sufficient here to speak of pragmatic reconstructions. (3) The requirement, already mentioned, to distinguish when speaking of 'transcendental' in the sense of a methodical beginning between '*tu-quoque*-arguments' and pragmatic (re-)constructions which are justified step by step.

Considering these reasons and considering as well the embarrassing variety of voices which have appealed to Kant in recent discussions of the concept of the transcendental, this concept cannot be applied, without reservations, to pragmatic reconstructions. A pragmatic reconstruction becomes no more intelligible by saying that it is transcendental. On the other hand, with respect to those parts of Kant's theory which I have discussed here (transcendental aesthetic, transcendental schematism), the statement is justified that pragmatic reconstructions take over the role of transcendental analyses. With that, not only the historical relationship between both conceptions would remain present; it could also be shown that pragmatic reconstructions realize, in a methodologically clear and terminologically uncompromised manner, certain intentions of Kant which are terminologically linked to the concept of the transcendental. To point out a *lebensweltliches Apriori* through such reconstructions is to make a 'transcendental argument' in foundational discourses about actual theories, since theories cannot 'prove' what they are pragmatically 'based upon'.

This transformation of transcendental analysis into pragmatic reconstructions may also be relevant to Kant's concept of 'transcendental deduction', whose results are 'transcendental concepts' (categories) and 'transcendental principles'. This deduction too is oriented towards a *factum*, namely the existence of conceptual knowledge which, according to our presentation, always refers to a non-conceptual *a priori*. Pragmatic reconstructions explicate this *a priori*, as has been shown in light of Kant's discussion of schematism. At the same time, the problematic relationship between genesis and validity which Kant attempts to clarify through an explication of 'transcendental conditions' (the 'actual' formation of knowledge stands under conditions which, at the same time, guarantee its validity) loses its apparent circularity. The 'foundations' of so-called transcendental concepts and principles are at issue, not their proof. Their foundation is a *factum*, i.e., a definite knowledge only in consequence of which it actually becomes meaningful to speak about 'conditions of possibility'; actually their foundation is also an *a priori*, i.e., an *a priori* extended by its pre-theoretical and non-conceptual elements. Only pragmatic reconstructions are in a position to explain both the *factum* and *a priori*-elements of knowledge.[54]

It is, therefore, obvious that the proposed transformation of transcendental analyses into pragmatic reconstructions, in the framework of the Kantian distinction between a 'synthetic-progressive' method and an 'analytic-regressive' method, gives preference to the 'analytic-regressive' method and, at the same time, throws metatheoretical considerations into relief. Reconstructions are always reconstructions of given knowledge, being first oriented towards its 'theoretical' form. Such a shift is consistent throughout with Kant's intentions as was emphasized at the beginning by pointing out that in Kant's analyses both methods actually fit into each other. The emphasis on a metatheoretical point of view, on the other hand, can be justified because, as a rule, '*a priori*' in Kant signifies '*a priori*-part of empirical-physical knowledge'. Furthermore, the explication of the concept of experience essentially follows this meaning. The elaboration of a distinction-*a priori* or the 'transcendental' status of predication established in this elaboration may, of course, serve as evidence for the possibility of a 'synthetic-progressive' procedure, now as before in existence, and that pragmatic reconstructions are not confined to theories.

These states of affairs once more show the systematic adequacy of the concept of (pragmatic) reconstruction concerning Kant's intentions in his use of the concept of the transcendental. It is important, however, that the concept of reconstruction is pragmatically established, i.e., that it is established not only on the level of propositions and conceptual explication (as defined by the concept of logical or rational reconstruction common since Carnap). Only in this way can the concept of reconstruction be linked to the idea of a justified construction. Only in this way too can those elements of a conceptual context be grasped within actual reconstructions which, like 'schematized' acts or intuitive constructions in mathematical matters, are themselves *not conceptual*, but nevertheless *constitutive* of the reconstructed context.

Since we have based our considerations on Kant, it might not be inappropriate to conclude in the form of an imperative: Establish within your linguistic and scientific constructions as well as within your linguistic and scientific *re*constructions a constitutive relation whose basis is always formed by a *lebensweltliches Apriori* (confined to pragmatic structures).[55] I shall call this imperative which, in the form of pragmatic reconstructions, also preserves Kant's idea of the transcendental, the *imperative of reconstruction*.

NOTES

[1] An excellent and much more detailed survey of the recent discussion can be found in R. Aschenberg, "Über transzendentale Argumente. Orientierung in einer Diskussion zu Kant und Strawson", *Philosophisches Jahrbuch* **85** (1978), pp. 331–358. New important contributions can also be found in P. Bieri and R.-P. Horstmann and L. Krüger (eds.), *Transcendental Arguments and Science. Essays in Epistemology*, Dordrecht and Boston and London 1979 (contributions by M. Baum, J. Bennett, R. Rorty, J. F. Rosenberg *et al.*).

[2] Cf. J. Passmore, *Philosophical Reasoning*, London [3]1973, pp. 58–80 (IV Self-Refutation). A pre-suppositional approach is advocated, for instance, by P. A. Crawford, "Kant's Theory of Philosophical Proof", *Kant-Studien* **53** (1961/1962), pp. 257–268; M. S. Gram, *Kant, Ontology and the a priori*, Evanston 1968; "Transcendental Arguments", *Noûs* **5** (1971), pp. 15–26; A. P. Griffiths, "Transcendental Arguments", *Proceedings of the Aristotelian Society Suppl.* **43** (1969), pp. 165–180; H. L. Ruf, "Transcendental Logic. An Essay on Critical Metaphysics", *Man and World* **2** (1969), pp. 38–64.

[3] *Critique of Pure Reason* B 740 (I use here Norman Kemp Smith's translation: *Immanuel Kant's Critique of Pure Reason*, London 1929, 1933, 1982).

[4] *Critique of Pure Reason* B XXII.

[5] Critique of Pure Reason B 735.

[6] J. Hintikka, "Transcendental Arguments: Genuine and Spurious", Noûs **6** (1972), p. 278; cf. *Logic, Language-Games, and Information. Kantian Themes in the Philosophy of Logic*, Oxford 1973, pp. 98ff.

[7] W. Stegmüller, *Probleme and Resultate der Wissenschaftstheorie und Analytischen Philosophie* IV/1 (*Personelle und Statistische Wahrscheinlichkeit. Personelle Wahrscheinlichkeit und Rationale Entscheidung*), Berlin and Heidelberg and New York 1973, pp. 39ff.; cf. "Gedanken über eine mögliche rationale Rekonstruktion von Kants Metaphysik der Erfahrung", I–II, *Ratio* **9** (1967), pp. 1–30, **10** (1968), pp. 1–31.

[8] S. Körner, "The Impossibility of Transcendental Deductions", *The Monist* **51** (1967), pp. 317–331, also in L. W. Beck (ed.), *Kant Studies Today*, La Salle Ill. 1969, pp. 230–244.

[9] *Critique of Pure Reason* B 519. In the same sense Kant also uses the expressions 'formal' or 'critical idealism', cf. *Critique of Pure Reason* B 519 footnote, *Prolegomena zu einer jeden künftigen Metaphysik, die als Wissenschaft wird auftreten können* §13 footnote III, §60 appendix.
[10] *Critique of Pure Reason* B 518f.
[11] *Reflexionen zur Metaphysik* 4642, *Akad.-Ausg.* XVII, p. 622 ("Die Verstandesbegriffe drücken alle *actus* der Gemüthskräfte aus, insofern nach ihren Allgemeinen Gesetzen Vorstellungen möglich sind, und zwar diese ihre Möglichkeit *a priori*").
[12] P. F. Strawson, *The Bounds of Sense. An Essay on Kant's Critique of Pure Reason*, London 1966, p. 18.
[13] *Critique of Pure Reason* B 25.
[14] For a consideration of the usage of 'transcendental' before Kant, see N. Hinske, *Kants Weg zur Transzendentalphilosophie. Der dreissigjährige Kant*, Stuttgart and Berlin and Cologne and Mainz 1970, pp. 20ff.; also I. Angelelli, "On the Origins of Kant's 'Transcendental'", *Kant-Studien* **63** (1972), pp. 117–122; "On 'Transcendental' Again", *Kant-Studien* **66** (1975), pp. 116–120. Further distinctions, applying mainly to the concept of the transcendental object, in G. Buchdahl, "Reduction – Realization: A Key to the Structure of Kant's Thought", *Philosophical Topics* **12** (1982), pp. 39–98.
[15] W. Stegmüller, *Probleme und Resultate der Wissenschaftstheorie und analytischen Philosophie* IV/1, p. 40.
[16] H. Heimsoeth, *Transzendentale Dialektik. Ein Kommentar zu Kants Kritik der reinen Vernunft* I (*Ideenlehre und Paralogismen*), Berlin 1966, p. 11.
[17] *Prolegomena* §§ 4, 5.
[18] Cf. G. H. v. Wright, *Causality and Determinism*, New York and London 1974; for the different meanings of 'condition' see F. Kambartel, "Bedingung", in *Enzyklopädie Philosophie und Wissenschaftstheorie* I, ed. J. Mittelstrass, Mannheim and Vienna and Zurich 1980, pp. 260–261.
[19] *Critique of Pure Reason* B 561.
[20] Cf. below pp. 92–94.
[21] *Critique of Pure Reason* A 118, cf. A 117.
[22] *Prolegomena* § 20 (I use here Lewis White Beck's translation: *Prolegomena to Any Future Metaphysics*, New York 1950).
[23] Cf. E. Schaper, "Arguing Transcendentally", *Kant-Studien* **63** (1972), pp. 101–116.
[24] *Critique of Pure Reason* B XVIII; cf. J. Hintikka, *Logic, Language-Games, and Information*, pp. 116f.
[25] *Critique of Pure Reason* B 197.
[26] The following argumentation is in line with my analysis of the conceptual history of '*a priori*': "Changing Concepts of the *a priori*", in R. E. Butts and J. Hintikka (eds.), *Historical and Philosophical Dimensions of Logic, Methodology and Philosophy of Science*, Dordrecht and Boston 1977 (*The University of Western Ontario Series in the Philosophy of Science* 12), pp. 113–128.
[27] *Critique of Pure Reason* B 2.
[28] *Critique of Pure Reason* B 11.
[29] Cf. F. Kambartel, "Wie abhängig ist die Physik von Erfahrung und Geschichte? Zur methodischen Ordung apriorischer und empirischer Elemente in der Naturwissenschaft", in K. Hübner and A. Menne (eds.), *Natur und Geschichte* (*X. Deutscher Kongress für Philosophie Kiel, 8.–12. Oktober 1972*), Hamburg 1973, pp. 154ff. G. Wolters has

pointed out the distinction between a proof-theoretical meaning and a foundational meaning of '*a priori*' in the thought of Kant's contemporary, J. H. Lambert. According to Wolters, a proof-theoretical *a priori*, viewed from Lambert's perspective, can be understood as a two-place predicate '*ap*'; an n-tuple $(S_1, \ldots, S_n)$ of propositions can be called '*a priori*' in regard to a proposition S, if S follows logically from $(S_1, \ldots, S_n)$:

$$(S_1, \ldots, S_n)\ ap\ S \equiv_{\mathrm{df}} (S_1, \ldots, S_n) \vDash S$$

Afterwards this proof-theoretical *a priori* is expanded 'foundationally' in Lambert, as in Kant, to include the concept of a measuring experience (G. Wolters, *Basis und Deduktion. Studien zur Entstehung und Bedeutung der Theorie der axiomatischen Methode bei J. H. Lambert* (1728–1777), Berlin and New York 1980, pp. 77ff.).

[30] *Critique of Pure Reason* B 80.

[31] For these distinctions see W. Kamlah and P. Lorenzen, *Logische Propädeutik oder Vorschule des vernünftigen Redens*, Mannheim 1967, pp. 212ff.; also P. Lorenzen and O. Schwemmer, *Konstruktive Logik, Ethik und Wissenschaftstheorie*, Mannheim and Vienna and Zurich [2]1975, pp. 210ff.

[32] "Der Raum ist kein empirischer Begriff, der von äusseren Erfahrungen abgezogen worden" (*Critique of Pure Reason* B 38).

[33] *Critique of Pure Reason* B 323.

[34] Cf. the relevant systematic considerations in P. Janich and J. Mittelstrass, "Raum", in H. Krings and H. M. Baumgartner and Ch. Wild (eds.), *Handbuch philosophischer Grundbegriffe* II, Munich 1973, pp. 1154–1168.

[35] Cf. F. Kambartel, *op. cit.* [Note 29], pp. 158ff.; also J. Mittelstrass, "Historische Analyse und konstruktive Begründung", in K. Lorenz (ed.), *Konstruktionen versus Positionen. Beiträge zur Diskussion um die Konstruktive Wissenschaftstheorie*, I–II, Berlin and New York 1978, II (*Allgemeine Wissenschaftstheorie*), pp. 256–277, here pp. 269ff.

[36] I recapitulate here, in a slightly modified form, my presentation in "Changing Concepts of the *a priori*" [above Note 26], pp. 120ff.; cf. also J. Mittelstrass, *Die Möglichkeit von Wissenschaft*, Frankfurt 1974, pp. 56–83, pp. 221–229 (Chap. 3: *Erfahrung und Begründung*).

[37] Another part, whose reconstruction refers to the experimental structure of empirical theories, is established by a *technical-a priori* (*Herstellungsapriori*) (for this cf. again the contributions mentioned in Note 36). In the case of physics this *a priori* is given by defining, in the framework of an *operational* theory, the geometrical concepts in terms of manufacturing procedures of spatial forms. A theory of geometrical determinations then already presupposes the manufacture of measuring instruments, whereby the *technical-a priori* leads to a *measuring-a priori* (*messtheoretisches Apriori*) of empirical theories like physics. Cf. P. Janich, "Zur Protophysik des Raumes", in G. Böhme (ed.), *Protophysik. Für und wider eine konstruktive Wissenschaftstheorie der Physik*, Frankfurt 1976, pp. 83–130.

[38] K. Lorenz, *Elemente der Sprachkritik. Eine Alternative zum Dogmatismus und Skeptizismus in der Analytischen Philosophie*, Frankfurt 1970, pp. 64ff.

[39] *Critique of Pure Reason* B 176ff.

[40] *Prolegomena* §7.

[41] For the following discussion cf. F. Kambartel, *Erfahrung und Struktur. Bausteine zu einer Kritik des Empirismus und Formalismus*, Frankfurt 1968, pp. 113ff.; also J. Hintik-

ka, "Kant on the Mathematical Method", *The Monist* **51** (1967), pp. 352–375; *Logic, Language-Games, and Information*, pp. 114ff.; A. T. Winterbourne, "Construction and the Role of Schematism in Kant's Philosophy of Mathematics", *Studies in History and Philosophy of Science* **12** (1981), pp. 33–46; and R. E. Butts, "Kant's Schemata as Semantical Rules", in L. W. Beck (ed.), *Kant Studies Today* [see Note 2], pp. 290–300; "Rules, Examples and Constructions. Kant's Theory of Mathematics", *Synthese* **47** (1981), pp. 257–288; *Kant and the Double Government Methodology. Supersensibility and Method in Kant's Philosophy of Science*, Dordrecht and Boston 1984 (*The University of Western Ontario Series in Philosophy of Science* 24), pp. 146–201 (Chap. VI: *Rules, Images and Constructions: Kant's Constructive Idealism*).

[42] *Critique of Pure Reason* B 75f.

[43] Cf. *Critique of Pure Reason* B 75.

[44] *Critique of Pure Reason* B 181.

[45] *Critique of Pure Reason* B 182.

[46] Cf. *Critique of Pure Reason* B 25; *Prolegomena* appendix A 204 note.

[47] R. Carnap, *Der logische Aufbau der Welt*, Berlin 1928, under the title: *Der logische Aufbau der Welt. Scheinprobleme in der Philosophie*, Hamburg ^{2}1961 (engl. *The Logical Structure of the World. Pseudoproblems in Philosophy*, Berkeley and London 1967).

[48] Cf. H. Reichenbach, *Experience and Prediction. An Analysis of the Foundations and the Structure of Knowledge*, Chicago and London 1938, pp. 7f.

[49] Cf. W. Stegmüller, *Probleme und Resultate der Wissenschaftstheorie und Analytischen Philosophie* IV/1, pp. 13f. For the history of the concept of reconstruction in recent philosophy of science cf. J. Mittelstrass, "Rationale Rekonstruktion der Wissenschaftsgeschichte", in P. Janich (ed.), *Wissenschaftstheorie und Wissenschaftsforschung*, Munich 1981, pp. 89–111, pp. 137–148; also (using some material from this article) "What does 'Reconstruction' Mean in the Analysis of Science and Its History?", *Communication & Cognition* **13** (1980), pp. 223–236.

[50] Cf. the presentation by P. Lorenzen in P. Lorenzen and O. Schwemmer, *Konstruktive Logik, Ethik und Wissenschaftstheorie*, pp. 181ff.

[51] On this point see F. Kambartel, "Pragmatic Reconstruction, as Exemplified by an Understanding of Arithmetics", *Communication & Cognition* **13** (1980), pp. 173–182. By the way, this notion does not imply the conception of 'transcendental pragmatics' or 'universal pragmatics' held, also with reference to Kant, by Habermas and Apel, which passes widely beyond the framework of philosophy of science used here. The essence of such a conception is a semiotic transformation of Kant's philosophy of subjectivity into a 'transcendental pragmatics of language', which, at the same time, is supposed to take into account the dependence of universal rules of communicative action upon contingent 'natural', i.e. historical-genetically determined, boundary conditions. Cf. K.-O. Apel, *Transformation der Philosophie*, I–II, Frankfurt 1973, II (*Das Apriori der Kommunikationsgemeinschaft*), pp. 155ff. (several relevant contributions under the skeleton title: *Transformation der Transzendentalphilosophie*); J. Habermas, "Vorbereitende Bemerkungen zu einer Theorie der kommunikativen Kompetenz", in J. Habermas and N. Luhmann, *Theorie der Gesellschaft oder Sozialtechnologie – Was leistet die Systemforschung?*, Frankfurt 1971, pp. 101–141; *Legitimationsprobleme im Spätkapitalismus*, Frankfurt 1973, pp. 152ff. (note 160); "Was heisst Universalpragmatik?", in K.-O. Apel (ed.), *Sprachpragmatik und Philosophie*, Frankfurt 1976, pp. 174–272. As against Apel, and intending a more intensive consideration of boundary conditions, Habermas argues

for a "minimalist interpretation of the transcendental" ("Was heisst Universalpragmatik?", *op. cit.*, pp. 198ff.). For a critical account of the proof-theoretical conceptions linked to 'transcendental pragmatics' or 'universal pragmatics' cf. C. F. Gethmann and R. Hegselmann, "Das Problem der Begründung zwischen Dezisionismus und Fundamentalismus", *Zeitschrift für allgemeine Wissenschaftstheorie* **8** (1977), pp. 342–368, here pp. 346–356.

[52] Cf. above Note 37.

[53] In this context Bubner explains the meaning of 'transcendental' according to Kant as the "demonstration of the lack of alternatives" of a knowledge which is reconstructed by a transcendental analysis (R. Bubner, "Zur Struktur eines transzendentalen Arguments", *Kant-Studien* [special isssue] **65** [1974], pp. 15*–27*, here p. 24*). The connection between an analysis of transcendental conditions and the possibility of competing theoretical approaches was established by S. Körner ("The Impossibility of Transcendental Deductions" [see Note 8]) and R. Rorty ("Transcendental Arguments, Self-Reference and Pragmatism", in *Transcendental Arguments, and Science* [see Note 1], pp. 77–103), but this point is problematic, at least within the framework of Kant-interpretation. For a critical account on this connection cf. D. Henrich, "Challenger or Competitor? On Rorty's Account of Transcendental Strategies", in *Transcendental Arguments and Science* [see Note 1], pp. 113–120, here p. 115).

[54] For a critical account cf. K. Lorenz, "The Concept of Science. Some Remarks on the Methodological Issue 'Construction' versus 'Description' in the Philosophy of Science", in *Transcendental Arguments and Science* [see Note 1], pp. 177–190 ("problems of justification are treated as if they were problems of constitution, i.e. as if they concerned investigations into the conditions of possible experience", p. 180).

[55] To explain further: With regard to the distinction-*a priori* it follows from this imperative, for instance, that terminological systems must be traceable, in a finite number of steps, to *ostensive determinations*, i.e. to predicates introducted by examples. Otherwise, there would not be, in fact, any explanation of predicates – because of the infinite terminological regress that would be unavoidable here.

SECTION II

CONSTRUCTIVISM AND PROTOSCIENCE

CHRISTIAN THIEL

PHILOSOPHY AND THE PROBLEM OF THE FOUNDATIONS OF MATHEMATICS[1]

On the 1st of May 1915 the philosopher Ludwig Wittgenstein wrote in his diary the following sentence: 'Not to occupy himself with problems which do not concern him is one of the principal skills of the philosopher'.[2] But does this not also apply to the problems of the individual sciences and especially to those of mathematics? Wittgenstein does not seem to have been of this opinion, at least not with regard to the foundations of mathematics; at that time and for years to come he occupied himself intensively with the foundations of mathematics. Since he did not, however, inform us of his reasons for so doing, I would like, once again, to make the question of whether the basic principles of mathematics are a legitimate subject for philosophical investigation the subject of my own deliberations.

A large number of both mathematicians and philosophers would surely wish to settle the issue of their respective competence as follows. It is the job of the philosopher to deal with the nature of the validity of mathematical propositions. It is the business of the mathematician, on the other hand, to determine the actual validity or non-validity of certain such propositions. According to this view it would be impermissible interference for the philosopher to criticize individual mathematical propositions or even to criticize the methods by which the mathematician obtains such propositions.

However, it is no secret that at the present time – and for the last ninety years or so – philosophical arguments have been employed to cast doubt not only on the validity of certain mathematical propositions but also on the permissibility of certain mathematical procedures. How could such a situation arise, bearing in mind that, for hundreds of years, mathematics had been regarded as a model of rigorous and almost infallible scientific method? Had it not indeed become the goal of the other sciences to achieve a rigour comparable to that of Euclid by imitation of the deductive procedures that he had employed? Philosophy was no exception, as is demonstrated by Spinoza's (admittedly unsuccessful) attempt at an ethics "more geometrico". And Kant, too, wrote in the Preface to the second edition of his *Critique of Pure*

Robert E. Butts and James Robert Brown (eds.), Constructivism and Science, 105–126.

Reason: 'From the earliest times to which we can trace back the history of human reason, mathematics has trodden the sure path of a science.' (B XI).

If the certainty of certain mathematical statements has, in the meantime, become a problem, then this is because, towards the end of the last century, this unique status of mathematics was suddenly shaken. Without an understanding of this crisis it is impossible to comprehend the present position of research into the foundations of mathematics and its relation to philosophy. I must, therefore, although my concerns here are primarily systematic, provide a brief and critical recapitulation of the origin of this crisis regarding the foundations of mathematics.

Up to the beginning of the 19th century it was usual, in analysis, to regard the propositions of the calculus as statements about infinitely small quantitites – as the name "infinitesimal calculus" reminds us even today. Cauchy, who had realised the unreliability of such alleged intuitions of the infinitesimally small, was able, in 1820, to replace all statements about infinitely small quantities by more comprehensible statements about limits. When, around the middle of the century, it was attempted to reduce the concept of the real number to more simple notions, it became evident that the architecture of analysis required, on the one hand, the rational numbers and on the other hand, either the concept of sequence or the concept of set. Georg Cantor suggested basing analysis (as a theory concerned with particular sets) on a general set theory which he had developed. The latter, however, relied so heavily on notions of the infinitely large that Cantor's colleagues rather tended to regard this theory, as far as its precision and reliability were concerned, as a retrograde step. It was not until towards the end of the century that Gottlob Frege presented an axiomatized system of logic which displayed a hitherto unknown degree of precision and which seemed suitable for providing a rigorous foundation for Cantor's theory of sets.

It is not without irony that it was precisely this precision which enabled Bertrand Russell to deduce a contradiction within Frege's system so stringently that there could no longer be any doubt of its collapse, nor of the collapse of Cantor's theory and also unfortunately of classical analysis. I am speaking here of Bertrand Russell's discovery of the paradox which is named after him, as published in 1903 by Frege himself. Further contradictions were discovered in quick succession and it is to these unpleasant discoveries that the shaking of the foundations of mathematics which I mentioned earlier can be traced back. Since this time there has existed the problem of the foundations of set theory and, since analysis cannot dispense with the theory of sets, a problem concerning the foundations of analysis and of all those mathematical theories constructed on analysis. These disciplines, which were quite crucial areas of mathematics, had to be put on a new and more trustworthy basis.

The way in which this might be done was and continues to be a source of dispute. But now the problem of the foundations of mathematics, expressed in this form, is a methodological one and thus the question of the sphere of competence of philosophy begins to appear in a different light. For even in the case of mathematics philosophers do not have to seize possession of methodological problems – perhaps by intimating that mathematics is much too difficult a science to be left to mathematicians alone. The position is rather that mathematicians themselves have long been in the habit of leaving problems of methodology to philosophers. Whether they were well advised to do so is, admittedly, another question. It is an unfortunate fact that many philosophers scarcely proficient in mathematics have advanced opinions about the foundations of mathematics at a level which has caused the mathematicians to no longer take any notice of their answers. On the other hand contemporary mathematicians are far from having attained the sense of methodological responsibility displayed by Frege or Russell and the situation has remained unchanged even in the face of the institutionalization of an autonomous branch of research into the foundations of mathematics, conceiving itself as a special mathematical discipline. As meta-mathematics, this discipline makes the propositions and procedures of other mathematical disciplines the subject of its research; but it has long since ceased to reflect on its *own* fundamental principles.

This, however, only shows that contemporary research into the foundations of mathematics has forgotten its origins and has lost sight of the problem which called it into being at the beginning of the present century. It was intended to resolve the crisis of confidence in mathematics by placing classical analysis and, as a first step to this end, general set theory on a secure basis. Eminent mathematicians like Hilbert, Brouwer and Weyl dedicated decades of their lives to the solution of this problem, fully conscious of the fact that they were, in the final analysis, struggling with an epistemological problem. In doing so they were guided by ideas which were ordered in something close to the following sequence of steps. As all mathematical disciplines employ logical deduction, the foundations of logic should be established first of all. If it should then prove possible to reduce arithmetic and the theory of sets to logic, as Frege had attempted and Russell, Carnap and others long after him, then their foundations would already have been established. Should such a reduction prove not to be possible, then an independent foundation would have to be found for them. With this the main problem would have been solved, for as is well known, analysis can be based on arithmetic and the theory of sets, and on analysis, in turn, can be based in the manner of analytic geometry, Euclidean geometry and finally, either once again analytically or with reference to Euclidean models, the non-Euclidean geometries.

It soon became clear, however, that mathematicians and those involved in research into the foundations of mathematics imposed widely differing requirements upon what such a foundation must achieve. If one ignores the neopositivists, who had moved to philosophy from the natural sciences, and certain isolated individuals like Hugo Dingler, then it must be said that philosophy has failed miserably even on this, its most traditional territory. In order to make clear the nature of this inadequacy on the part of philosophy and the responsibility which it still bears with respect to the problem of the foundations of mathematics, I shall proceed to outline and to comment critically upon three contemporary attempts at establishing the foundations of mathematics. At the same time I shall venture an appraisal of the situation in which the controversy over the foundations of mathematics finds itself at the present time.

The first suggestion with regard to establishing the foundations of mathematics is of a type essentially derived from problems relating to the foundations of physics. It was employed as early as 1690 by Christian Huygens in his '*Traité de la lumière*' in which he declared that he would justify the underlying principles of this work in a manner different from that usual in mathematics; namely, by the success of conclusions which could be drawn from them. This procedure must have rapidly met with the approval of mathematicians for, as early as 1720, in 'The Analyst', his paper critical of the foundations of mathematics and directed against the adherents of the Newtonian differential calculus, George Berkeley employs the argument that in a mathematical theory the conclusions would always have to be established on the basis of the principles and not the principles on the basis of the conclusions.

This suggestion is of course incomplete until it is stated just how the conclusions can be justified. This is to happen by their standing the test when applied practically, whereby predominantly physical applications are to be thought of – this, at least, is the opinion of those who support this suggestion. According to this view mathematical propositions which manifestly prove themselves in physics and technology can be neither senseless nor false. It is much more the case that their having confirmed themselves in practice speaks for their correctness or, more precisely, that their correctness is to be understood as consisting precisely in this confirmation. For this reason theories which are accessible to confirmation have been designated as 'application theories'. The suggestion outlined above, however, as the 'prognostic foundation' since the decision as to whether an application theory has indeed been shown to hold good depends on whether its propositions can also be demonstrated to be valid in the future. In this sense even the theory of sets would be an (admittedly very general) system of physical hypotheses.[3]

This attempt at a prognostic proof has tangible weaknesses. Mathematicians themselves are by no means content with the requirement that their propositions should hold good in the sphere of physical application. Even natural scientists do not wish to test a mathematical proposition through its application. If there are mixed together a litre of each of two different liquids and the result is not exactly two litres then the experimenter will not regard the arithmetical proposition $1 + 1 = 2$ as having been proved false and request a new arithmetical theory from his colleagues. He will look for reasons for the unexpected disappearance phenomenon. Mathematical presuppositions are, therefore, taken as given in the testing of physical or chemical hypotheses. Thus the idea that the validity of mathematical propositions is prognostically based may, indeed, cause certain persons to have increased faith in mathematics but it cannot be seriously considered as an approach to the problem of foundations, nor will it be defended as such by mathematicians themselves.

The most widespread view held today is that of an axiomatic foundation of mathematics. I shall concern myself more closely with the latter, although the subject cannot possibly be done adequate justice in the short space of an article. The axiomatic or deductive process is based on the notion of accounting for and rendering comprehensible the totality of true propositions of a science by taking a finite and smallest possible number of propositions to form the basis of this science – which are then termed principles or axioms – while all other true propositions are derived from them by logical deduction.

This idea must have been arrived at very early, for the classical axiomatic presentation of ancient mathematics in Euclid is already the culmination of a longer period of development, which we are, unfortunately, able to reconstruct only inadequately. There could be no possible objection to an axiomatic construction and it would indeed considerably simplify the problem of foundations were we to succeed in determining both the truth of the finite number of basic axioms and the correctness of the deductive procedures employed. The extent to which Euclid attempted this is controversial but a later (5th century) commentator, Proclus, was the first to express a different view of the axiomatic process, which is the dominant one today and whose usefulness in respect of the foundational problems one can legitimately doubt. For Proclus writes: '"The writer of an elementary text book of geometry (must) teach the principles of a science separately and the conclusions drawn from the principles separately; he need not render an account of the principles but of the conclusions drawn from them he must"'.[4]

Whether Proclus had ideas similar to those of the contemporary proponents of an axiomatic foundation of mathematics can scarcely be determined, but this is comparatively unimportant with regard to foundational problems widely prevalent today. In order to elucidate this I must once again briefly return to the historical development. The above-mentioned shaking of mathematics by the antinomies based on the theory of sets did not, of course, have the effect of bringing about a temporary stop in the teaching of and research into mathematics and a suspension of the application of mathematical propositions until such time as a satisfactory solution of the foundational problem would have been found. Hence the first aim was to construct a system of analysis which could at least be demonstrated to exclude the known contradictions. Ernst Zermelo made an important step in this direction by constructing a system of axioms for general set theory in 1908 which preserved Cantor's intentions while ensuring that the contradictions which had occurred hitherto were not deducible from it.

Zermelo himself was clear that this could not suffice. Who could say whether other antinomies might not be deducible from his system? Zermelo, therefore, taking up a more general Hilbertian requisition demanded for the justification of his axiomatic theory of sets a proof of consistency. Researchers into the foundations of mathematics have today become accustomed to see this demand as containing everything which can be required of a rational foundation of an axiomatic system. Zermelo never deceived himself about the fact that the consistency proof was a minimal demand; he had also honestly added that he would separate off "the question of the origin and domain of validity" of the axioms as "more philosophical". He did not thereby maintain that an answer to this question would in the future be superfluous. Indeed, the situation is such that there has hitherto been no successful consistency proof either for Zermelo's original system or for the more recent, improved version – the Zermelo–Fraenkel (ZF) theory of sets which is regarded highly by contemporary mathematicians. Zermelo's requirement, therefore, still remains to be met. Historically, the demand for a consistency proof became the nucleus of the Hilbert Programme for a new foundation of mathematics. This programme, which in its details is considerably complex, regards mathematical theories in their axiomatized form as a game of deduction in which from certain initial positions other positions are derivable according to given rules. The task is to prove that in such a game it is never possible to obtain two playing positions which can be interpreted with regard to their contents, as a certain mathematical proposition and its negation. Frege had already had this idea but he did not carry it through because he thought that he could supply a consistency proof for his system in a more simple way. Here he was, admittedly, wrong as is demonstrated by the Russell paradox. It was now Hilbert and his fellow workers in Göttingen – that stronghold of mathematics – who set about trying to supply a consistency proof for classical analysis in the way indicated.

Zermelo's approach was developed into a position in the theory of foundations which imposed certain requirements on the means that could be used to obtain a consistency proof. These methods of proof, even though applied at a so-called meta-level, must yet be seen as involving deductive procedures and concept-formations of a strictly mathematical nature. Thus demonstrating the reliability of the entirety of mathematics in the sense of the Hilbert programme also involves proving the reliability of these methods. This enterprise may be regarded as circular. The ideas of the Hilbert school were along the lines that mathematical methods must indeed be applied in the proof, though only those formulable within the system under examination itself. They should on the other hand be reliable, more reliable at least than the axiomatic system for which a consistency proof is sought.

It here becomes clear that this so-called "formalist approach" assumes a quite peculiar relationship of foundation between mathematics and meta-mathematics. One could only speak in any clear sense of a proof with regard to the relation described above, if the arguments used in the proof of consistency of an axiomatic system are based on weaker assumptions than those of the system itself. It was therefore a heavy blow for the Hilbert programme when Kurt Gödel discovered in 1931 that a consistency proof for an axiomatic system which contains, at least, arithmetic could never be obtained with means formulable within this system. In other words *stronger* means are required to prove the consistency of such a system than can be formulated within it. Thoralf Skolem made the following very apposite remark on this point: ‘“This is as if we should hang up the ground floor of a building to the first floor, this again to the second floor, etc.”’[5] Obviously, this approach results in a situation almost diametrically opposed to the spirit of foundational claims. One could therefore rightly speak of the failure of the Hilbert programme.

But what in fact had been disproved? Hilbert himself had differentiated between the axiomatic construction of a theory and the ‘necessity of providing a foundation for the propositions underlying the theory as axioms’. His suggestion for such a foundation was to limit oneself to axiomatic construction. When he also supplied an intuitive foundation for elementary arithmetic in 1923, he said he had only chosen to go this way for the sake of brevity and that in the final version of his theory he would supply an axiomatic foundation also for arithmetic. Hilbert's aim was as he formulated it ‘to penetrate ever deeper layers of axioms.’

To me Hilbert's position has always been puzzling because it contradicts all of the foundational notions customarily adhered to in the most diverse branches of the philosophy of science. Consider the status of a proposition A deduced from a proposition B, whose foundation is, in turn, only possible by application of proposition A,

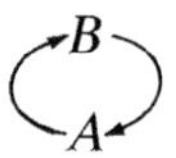

(such circular proofs can appear in very subtle forms – in impredicative notions, for example). It would seem surely that if A, here, cannot be regarded as having a foundation, then neither can a proposition A

which someone might derive from a proposition B, the latter in turn from a proposition C, followed by the declaration, 'and so on.'

$$\begin{array}{rr} & A \\ & B \rightarrow A \\ C \rightarrow B & \\ \cdot & \\ \cdot & \\ \cdot & \end{array}$$

At some point a halt must be called at a proposition which has to be believed, in the absence of a foundation. We can be sure in such a case that our collocutor, who may be safely supposed to have only a finite number of years of life behind him, has not gained his apparent insight into the validity of this proposition by such an infinite means. And thus we may properly deny to him the acknowledgment of such an argument as proof.

While the Gödel result of 1931 (mentioned above) demonstrates that the foundational programme sketched by the Hilbert school is impracticable, a further discovery of Gödel's reveals the limitedness of the axiomatic method in general. He showed that every non-contradictory system of axioms with sufficient expressive power is incomplete in the sense that it is always possible to make true statements within the relevant theory which are not deducible from the given system of axioms. These limits cannot, in principle, be overstepped. Even when the original system of axioms is complemented by the Gödelian statements not derivable from it, new, true but undeducible statements of the theory can be specified for the thus extended system of axioms. This is repeated with every step in the process of extension. Even the Peano-system is incomplete in this sense and would, therefore, even in the event of a successful proof of consistency in Hilbert's sense, be unable to provide a foundation for the whole of arithmetic.

In his later work Hilbert revised his ideas with regard to the foundation of mathematics under the influence of Gödel's findings. He no longer makes the reliability of the means employed in a consistency proof dependent on its ability to be formulated within the system under examination. Instead, he now requires the *immediate* discernability of the appropriateness of these means in a sense similar to that required by the opposing party in the dispute about the foundations of mathematics, the intuitionists. I have so far refrained from discussing intuitionism since the philosophical background of this movement is scarcely more convincing than that of the ontologizing or platonizing movements. All the insights for which we must be grateful to intuitionism have long been adopted by the proponents of the third suggested proposal, to which I shall now turn – the weaknesses of the axiomatic attempt at foundations having hopefully become clear.

I would like to begin by saying that it is not that this third suggested proposal necessarily seeks to disallow the axiomatic construction of mathematical theories. Quite the contrary. It does, however, proceed from a better understanding of the character of axiomatic theories. This understanding, which is also characteristic of the work of the philosopher (and ex-mathematician) Edmund Husserl, had already been adopted by Frege against Hilbert. Frege had, however, failed to make any impression on his distinguished colleague who proved himself very dogmatic on this point. This insight maintains that axiomatic theories are formal theories which do not represent foundations but only *forms of possible foundational connections* which would first have to be provided in individual cases. In the language of mathematicians, the models of axiomatic theories would first have to be constructed. If the contents of a mathematical discipline have been established in this way, then and only then can it be simplified, that is to say axiomatically constructed.

The fact that contemporary mathematics under the influence of the Bourbaki Circle, customarily dispenses with first establishing a constructive basis for mathematical theories has been subjected to bitter criticism from one philosophical camp (and on this occasion, from mathematically thoroughly competent quarters). It would mean, writes Friedrich Kambartel, for example, dispensing with substantial foundations relating to the content of mathematical propositions and hence 'in the final analysis with the foundations of propositions altogether.'[6]

What do those workers in foundations who are favoured by philosophers have to offer on the other hand? In order to supply acceptable foundations for those mathematical theories which continue to be axiomatically constructed they first draw conclusions from the failure of the Hilbert programme. It is no longer required of the means employed in a consistency proof that it should be formulable within the system under examination itself. There are, of course, proof procedures which may be regarded as being valid independently of any systems into which they can be incorporated, for example, mathematical induction. If I can demonstrate of a property $E(x)$ that it is true of the number one, and have a procedure which, transfers the property $E(x)$ from any positive whole number, to the subsequent number, I may then be sure that the property $E(x)$ is in fact shared by every positive whole number.

This is intuitively evident and it is transparently 'finite'. The formulation of this insight, the proposition designated as the 'principle of mathematical induction'

$$E(1) \wedge \bigwedge_m . E(m) \rightarrow E(m+1) \dot{\rightarrow} \bigwedge_n E(n),$$

may thus be held to be valid quite independently of whether I employ it to prove the freedom from contradiction of an axiomatic system in which this proposition is itself not expressible.

Similarly recognizable as reliable are means usually termed "combinatorial", although this designation is not particularly appropriate. Let us assume we have a primitive axiomatic system with the rule of substitution and the separation rule

$$\frac{\begin{matrix} S \\ S \rightarrow T \end{matrix}}{T}$$

for the derivation of propositions from the axiom system which consists of the two axioms $p \rightarrow p$ *and* $\neg (p \rightarrow p)$. Let 'p' be here an *eigenvariable* which may therefore only be replaced by expressions which have already been derived from the two above-mentioned rules. We can then see, by 'combinatorial' means, that the variable 'p' is not derivable singly. For each of the two axioms contains the variable 'p' twice. By application of the substitution rule, that is to say with each instance of its application, we continually obtain an even number of variables. The same holds for the application of the separation rule, since each derivable formula of our axiomatic theory (with the exception of the axioms themselves) contains, both before and after the arrow, variables of even number. The formula 'p' itself, however, contains the variable 'p' exactly once, that is, with the frequency of an odd number. If cannot, therefore, belong to the formulae deducible within our axiomatic system. This is not a naive example. It shows that, in contrast to the manner in which the topic is presented in many text books, it is not always possible to demonstrate the consistency of an axiomatic system by drawing attention to the presence of some correctly formed but non-derivable formula. 'p' is correctly formed but not deducible and the system is nevertheless contradictory, if we interpret 'p' as a propositional variable, $\neg$ as negation and $\rightarrow$ as implication.[7]

These questions of detail are, however, only intended to exemplify the nature of finite methods. At the level we are here concerned with, post-Hilbertian meta-mathematics is still indecisive in that no attempt is made to delimit the extent of the evidence to be regarded as reliable. This is readily understandable, since after the Hilbert programme, caution will be exercised in not setting premature limits to the meta-mathematical evidence. And there can of course be no hope whatsoever (at this stage) of being able to undertake a delimitation of what is reliable for all time; the domain of 'finite evidence' is itself indefinite.

There are, however, two reasons for at least attempting an initial delimitation. The first is technical: only with a formally comprehensible delimitation of the means of construction and proof can it be precisely stated how much of the (classical) mathematics that it is intended to construct and to justify can in fact be accounted for. The second reason is more philosophical: the procedures regarded as reliable should once more be examined with a view to determining whether or not they exhibit a common characteristic property, to which they own their especially reliable character.

It seems to me that the most valuable insights have been supplied by the position that goes by the name of 'operativism' or 'constructivism'. As with Hilbert's position, the objects under examination are the mathematical symbols – that is to say formulae, propositions and proofs. But these symbols and sequences of symbols are not only regarded, with regard to their form, as physical objects, but also as objects which we produce with certain intentions in mind and with which we undertake certain actions (with a particular intention). Hence, even the most general calculi are to be regarded as systems of rules with beginning rules $\Rightarrow A$ and continuation rules $B \Rightarrow C$, whose proof can only consist in the fact that acting in accordance with them serves a justified and expressly stated purpose. The advantage of this approach is clear. We have almost complete insight into purposeful action of this sort. We can in each case check whether the desired aim has been reached and can always establish whether an action in accordance with a suggested scheme of action, for example a rule of the calculus, has been successful or not. This then especially holds for the individual steps in deduction from a system of axioms which has been previously justified as a means for the expression of an already constructed mathematical theory.

The proponents of this operative position here like to quote Kant who once said that we can only really understand that which we have produced ourselves. The often accompanying remark to the effect that 'in mathematics there exists only that which we ourselves can produce' usually meets with massive resistance. Insofar as this rejection is built up on the belief in an ideal world in which the objects of mathematics already exist, I wish to ignore it here, since such a point of view would of course for its part require a defence. Where this dislike of the operative thesis is not based on a belief of this sort, it is normally rooted in a plausible but inappropriate analogy. In other sciences it is mostly not the case that we regard as existing only that which we ourselves or other people have produced. The zoologist correctly assumes that there are lions, even if lions are not produced by human beings. The situation is, however, already different in the humanities where we wish to understand texts and can understand them only because and to the extent that they have been produced on the basis of a mutual communication-system. In mathematics we are far closer to the humanities than we are to zoology. We at first produce, in principle, certain artifacts, for example numerals or figures of elementary geometrical forms, and then begin with the construction of a mathematical language about our actions with regard to these artifacts, about the aims of these actions, about their results, etc..

The illuminating and compelling manner in which those propositions which are usually placed at the beginning of a mathematical discipline as axioms (as unproved and unprovable) are able to be fully proved from this practical basis never fails to surprise. For constructive arithmetic the process can be outlined as follows.[8] We learn what counting means and to what end we count something from the practical experience of everyday life. We also know from this practical experience that counting is not dependent on the nature of what is being counted but that we can count everything for which we have a predicator, whereby we usually assign normal predicators (*individuativa*) individually, material or mass predicators (*continuativa* or *collectiva*) always in connection with words such as, for example, "a piece of ..."[9]

How often we have assigned a particular predicator to the individual units of a totality (always proceeding from one such unit to another as yet unselected one from the same totality), the measure of the manifoldness of our assignment of the predicator, can be established as if it were a 'track'. We can, for example, for every act of predication put a stone in a box or put a line on a piece of paper or employ counting words whose sequence has previously been fixed and then note the one which was the last to be uttered.

In contrast to speaking, counting is not part of the practical everyday experience of each normal person. It can, however, be learned by the acquisition of certain counting procedures, that is to say by learning a certain counting-process, a process of acting according to a certain scheme of action. Which special counting procedure is selected is unimportant. Counting (the 'measurement of the manifoldness of our predications') can, for example, be done by the setting up of so-called lists of marks. We here make with each predication a mark on some material or other, making a stroke on beginning to predicate according to the following beginning rule N_a:

$$N_a\colon \Rightarrow |.$$

We make, with each further predication, from the list of marks which we have so far produced (indicated here by the letter 'n' which plays the role of substitute for lists of marks produced by application of rules N_a and N_f), another list of marks according to the following continuation rule

$$N_f\colon n \Rightarrow n|,$$

by adding another mark to the right. N_a and N_f are thus rules for the production of figures which can serve as counting symbols and which we here prefer to other kinds of counting symbols because of their simplicity. We now establish a terminology by agreeing to call numerals all (and only the) figures produced according to the production rules N_a and N_f. The figure '|' is therefore to be called a numeral, and given a numeral 'n', the figure '$n|$' is also to be called a numeral:

$$\Rightarrow | \text{ is a numeral}$$
$$n \text{ is a numeral} \Rightarrow n| \text{ is a numeral}.$$

Such a terminological statement which establishes the appropriateness or non-appropriateness of a predicator (here 'numeral') for certain objects (here the figures), by referring in an immediately comprehensible way, to the production rules for the objects is called an *inductive definition* of the predicator.

Anyone agreeing with the above will, on the basis of it recognize the following statements as true:

Theorem 1: $|$ is a numeral.
Theorem 2: If n is a numeral, then $n|$ is a numeral.

We here use the predicator 'theorem' for 'true statement', not for 'statement deduced from the axioms'. The latter would, of course, be an as yet incomprehensible notion, as we have not yet introduced the expression 'axiom' nor have we had any reason for so doing. The propositions 1 and 2 can be recognised as true statements not on the basis of any 'experience' with figures constructed according to N_a and N_f but simply by once more reminding ourselves of the adopted terminology. Propositions of this sort, whose truth can be ascertained by recourse to terminological conventions alone, we shall call '*analytic propositions*'. Like all statements, such propositions can be asserted and it is sensible to say that they are 'true'. Analytic propositions do not prescribe a particular terminology but their truth is demonstrated by recourse to terminological conventions.

If one has learned to proceed according to N_a and N_f, one can also understand what it means to undertake equal steps in construction according to these rules. This means of course nothing more than repeating an action according to a scheme of action. Numerals that can be produced according to the rules N_a or N_f in equal steps in construction may be called 'equal'. There is obviously a decision procedure for the equality of two numerals. A step in the decision process consists in taking away a figure $|$ from each of the numerals, let us say the one furthest to the right ("reversal" of N_f). These steps are repeated until one of the numerals disappears, that is to say no figure $|$ is left over. If, after this last step the other numeral has also disappeared, then both numerals were equal (one has arrived at the 'beginning' of N_a), otherwise they were not. Hence the following rule-system provides a full specification of the true statements of equality:

$$G_a\colon \Rightarrow | = |$$
$$G_f\colon m = n \Rightarrow m| = n|$$

This is a terminological convention for the two-place predicator (relator) 'equals', which, analogous to our procedure with Propositions 1 and 2, provides the following analytically true propositions:

Theorem 3: $| = |$
Theorem 4: If $m = n$, then $m\,| = n\,|$

And further

Theorem 5: For no m is $m\,| = |$

If we subtract a figure from the right of both of the numerals occurring in an equation $n\,| = n$, then the numeral to the right of the equals sign disappears while the numeral on the left does not disappear (the numeral n remains). Thus according to our decision procedure the two numerals $n\,|$, $|$ were not equal.

Theorem 6: If $m\,| = n\,|$, then $m = n$.

If we apply our decision procedure to $m\,|$ and $n\,|$ we obtain after the first step, the subtraction of a figure from the right hand side of each of the two figures in the equation, precisely the numerals m and n. Hence, if the continuation of the procedure leads to the decision that $m\,|$ and $n\,|$ were equal, the steps taken subsequently at the same time furnish proof that the figures m and n are equal, for these steps are fully identical to the steps which lead to the decision 'equals' when the decision procedure is applied to the figures m and n. Finally, the following is valid:

Theorem 7: If the numeral $|$ has a property E and if this property is such that in all cases where it belongs to a numeral n, it also belongs to the numeral $n|$ derived from it according to the rule N_f, then each numeral shares the property E.

Let us assume, for example, that E is a property of the sort mentioned in the theorem and that m is a given but completely arbitrarily chosen numeral. According to the first assumption the numeral $|$ has the property E. If m is equal to the numeral $|$, then we must further consider that, according to the second assumption, the property E must also belong to the numeral || as the latter is derived from numeral | according to the rule N_f. If m is equal to the figure ||, then we have finished. Otherwise we further conclude that, according to the second assumption the numeral ||| also has the property E as it was derived from the numeral || according to the rule N_f.

As numerals can only be derived according to the rules N_a and N_f, we shall, by continuing our procedure, at some time or other arrive at some figure equal to the given numeral m, of which we have at the same time proved that it shares the property E. As m was chosen arbitrarily and our procedure ultimately accounts for each such figure, we have demonstrated that the property E in fact belongs to every numeral.

By supplying proof of the theorems 1,2,5,6,7 we have demonstrated that the so called Peano–Dedekind system of axioms of arithmetic,

(1) $| \in Z$
(2) $n \in Z \rightarrow n' \in Z$
(3) $\bigwedge_{m,n} . m,n \in Z \rightarrow m' = n' \rightarrow m = n.$
(4) $\bigwedge_m . m \in Z \rightarrow m' \neq 1.$
(5) $\bigwedge_m . m \in Z \dot{\rightarrow} E(m) \rightarrow E(m') \ddot{\rightarrow} E(1) \rightarrow \bigwedge_n E(n).$

when interpreted as follows becomes a system of true propositions. Let the domain of variability of the individual variables $m, n \ldots$ be the domain of the figures which can be derived from N_a and N_f, the individual constant 1 as the numeral |, $n \in Z$ as "n is a numeral", $m = n$ as equality of numerals ($m \neq n$ as the corresponding inequality), n' as n | in rule N_f. Let E be a schematic variable for properties constructively defined for numerals. Let $E(n)$ be the proposition expressing that the property E belongs to the numeral n. The symbols $\rightarrow$ and $\bigwedge$ are implication and universal quantifier, the punctuation being in accordance with the conventional rules for the construction of logically complex propositions (it goes without saying that we could also render our propositions 1 to 7 in logical symbols and could either have justified them according to the rules of dialogical logic or could have derived them from a logical calculus – this would merely have required more space).

Since this interpretation has recourse to the production of numerals (and thus to certain acts), we speak of an 'operative model' of the Peano–Dedekind system of axioms. This model is complete, in contradistinction to the above-mentioned system of axioms (conceived, when taken together with a logical calculus, as fully formalized). It is clear that the validity of all the hitherto justified propositions about numerals is not dependent on the mode of rendition of those numerals which we have chosen as counting-signs. In arithmetic we are also only interested in those statements about numerals whose validity is 'invariant' with respect to the replacement of the numerals occurring in them by others established by means of the same constructive steps (for example, by means of an instruction referring to our production rules, "N_a, N_f, N_f, N_f" for a numeral of the form "||||"). Instead of adding to the statement A $(m_i, \ldots, m_k)$ about the numerals $m_i, \ldots,$ m_k the complementary statement that this statement possesses the above-mentioned property of invariance

$$A(m_1, \ldots, m_k) \wedge \bigwedge_{n_1 \ldots n_k} . m_1 = n_1 \wedge \ldots \wedge m_k = n_k \cdot \rightarrow A (n_1, \ldots, n_k)$$

we no longer formulate A $(m_1, \ldots, m_k)$ as a statement about numerals but as a statement about cardinal numbers. Thus, for example, in regard to Theorem 5 the form "For no cardinal number m is $m \mid = \mid$", replaces "For no numeral m is $m \mid = \mid$". Since, in doing this, we are paying attention only to the steps of construction, but ignoring or 'disregarding' differences in the rendition of the numerals, we may here speak of an "abstraction step".

In the sense thus established we arrive at cardinal numbers (more precisely at the expression 'cardinal number' by 'abstraction'.[10] The propositions delimited in this way then form the content of arithmetic.

One can now obtain the first of what might be called "expansions of the concept of number", leading from the cardinal numbers to the positive rational numbers, by a further limitation. For, proceeding from the stage we have now reached, the positive rational numbers can be introduced by means of a further abstraction, by establishing for the pairs of numerals (m,n) an equivalence relation

$$(m,n) \sim (\bar{m},\bar{n}) \leftrightharpoons m \cdot \bar{n} = \bar{m} \cdot n$$

(addition and multiplication of numerals being explained inductively in the usual manner). Statements about pairs of numerals whose truth or falsity remains invariant in the face of the replacement of one of the pairs of numerals occurring in the statement by a pair of numerals equivalent to it can then be formulated as statements about positive rational numbers (m/n) so that we have

$$A(m/n) \leftrightharpoons \bigwedge_{(x,y)} . (m,n) \sim (x,y) \rightarrow A(x,y).$$

The domain of numbers thus obtained can now be extended to the full domain of rational numbers by the introduction of zero and of the negative rational numbers by making, for all statements in which these new numbers occur, equivalent statements, in which, as was hitherto the case, only positive rational numbers occur. If one then finally defines an equivalence relation for statement-forms about rational numbers by means of

$$A(\in_x B(x)) \leftrightharpoons \bigwedge_c . B(x) \tilde{x} C(x) \rightarrow A(C(x)) .,$$

one can then introduce "sets of rational numbers" through a further process of abstraction by formulating statements whose truth or falsehood is invariant to the replacement of one of the statement-forms occurring in them by another equivalent to it as statements about 'sets of rational numbers'.

With the assistance of these sets a constructive analysis can be built up by subjecting the statement-forms employed in the abstraction of sets to appropriately strict requirements. With each permissible replacement there must result a statement whose sense is already fully explained without recourse in the resulting statements to the statement-forms in question, or even to the set abstracted from it and without of course employing unlimited quantification over sets. The so-called impredicative concepts (see p. 112), whose role in the generation of antinomies had become clear, are thus excluded from the outset.

After the surprisingly simple establishment of a constructive arithmetic it is even more remarkable what progress can be made by the extension of constructive procedures to a foundation of analysis. It has, after all, been possible on the basis of constructive mathematics to provide a foundation for logic; a consistency proof for arithmetic; and even to demonstrate the consistency of that part of classical analysis which is in fact utilized in all applications. Hence, while the Hilbert programme may not actually have been completed, its goal has in fact been reached.

The successful realization of this aim (around the year 1950) represents a quite substantial advance on the mere propagation of constructive methods characteristic of earlier intuitionism. Nevertheless the objections which were then made to intuitionism by the mathematical establishment are still made today with respect to attempts at a constructive foundation of mathematics. Paul Bernays, for example, a distinguished colleague of Hilbert's, wrote of intuitionism: "It would appear that philosophy is making unreasonable demands on mathematics in claiming that it should give up its more simple and productive method in favour of a method which is clumsy and less systematic without having any inner compulsion for doing so. This approach makes the standpoint of the intuitionists suspect."[11] A similar objection directed against constructivism today would be completely antiquated. The justification of constructive logic on the basis of transparent dialogue rules by Paul Lorenzen and Kuno Lorenz (since about 1960) has not only, for the first time, permitted a construction of the customary logical calculi which really deserves to be called a foundation for the same; it has also given rise to a foundation of arithmetic and analysis which can no longer properly be designated as 'clumsy'. Finally, constructivism has offered a constructive theory of sets in which the choice of axioms is no longer left to intuition or merely to authority, as was the case with the construction of Zermelo's system on the basis of Cantor's intuitive theory of sets.

I must ask the reader to excuse the digressions which I have imposed on him in this paper. I have made an excursion into the history of modern mathematics and its foundations in order to elucidate a development whose findings have, as yet, not been as well received as an understanding of the contemporary situation would demand. I have made a further excursion – without being able to go into the necessary detail – in order to provide some information about one suggested proposal for resolving the dispute concerning foundations, a proposal which I regard as being the best available at the present time. I would ask for the reader's kind forbearance if, contrary to my intention in providing this information, my exposition has taken on the character of a somewhat pedagogical lecture. I should like, after these digressions, to bring my remarks to a close by returning to their systematic aim.

What role, after what has been said, can be conceded to philosophy in respect of the problem of the foundations of mathematics?

If my previous estimation of the situation is correct, we have, in the attempt at a constructive justification of mathematics, a basic point of view which is able in a convincing manner to justify its claims with regard to foundations. 'In a convincing manner' means, in this case, not a noncommitted appeal to the alleged evidence of the basic propositions of mathematics, but rather binding assertions about the derivability and underivability of figures in calculi, about the eliminability of steps in proofs, and about the winning of dialogues about logically constructed mathematical propositions. Even if the problem of the foundations of mathematics is as yet by no means completely solved – geometry above all contains a series of hard nuts to crack – the present approach has now proved itself, over more than thirty years, to be exceptionally fruitful. It is only recently that Lorenzen has produced a consistency proof which should remove the doubts of contemporary meta-mathematicians about finally putting arithmetic on a secure basis, a proof which has a simplicity of which Hilbert would not even have dared to dream.

If the problem of the foundations of mathematics has thus approached a technical solution or has indeed been solved, then we can say that for the first time in this century, as far as I can see, philosophical requirements have been adequately satisfied. This is of course no mere accident. It is rather the result of the efforts of a small group of mathematicians who have gone over to the camp of the philosophers – some to the extent of changing their subject – in order to work on a foundational project whose rigour would exceed even that of Euclid. They have suggested a foundation which, without sacrificing any of its declared aims, is also able to hold its own in the face of the socially critical reflection demanded by mathematicians from the point of view of moral argumentation in the dispute over the foundations of mathematics, or of what we might call the ethics of mathematics.

As a member of an Institute of Philosophy I would prefer not to close by suggesting the indispensability of philosophers to mathematicians. I would prefer to finish with a reference to the almost inestimable benefit which the crisis in mathematics and the problems of its foundations have brought for philosophy. Until well into the nineteenth century distinguished physicists and other natural scientists understood their fundamental research as philosophy. This was no mere quirk on their part, it reflected the view that the construction of a methodological foundation of a scientific system of propositions was considered to be a natural task of philosophy. It is through the crisis in mathematics that 20th-century philosophy became once more conscious of this task. I would like to hope that fruitful cooperation of basic research in mathematics and philosophy will not only provide documentary proof, available to the other sciences, of the usefulness of philosophical reflexion, but that it has also confronted philosophers with a possible field for practical philosophizing of which they had almost completely lost sight.

NOTES

[1] Revised and enlarged version of a lecture given at the Technical University in Aachen on the 9th June 1971. It was first published in Spanish translation (by Professor J. Sanmartín Esplugues) under the title "El Problema de la Fundamentación de la Matemática y la Filosofía" in *Teorema* **3** (September 1971), pp. 5–24.

[2] Ludwig Wittgenstein, *Notebooks 1914–16*, Oxford, 1979 (2nd ed.), p. 44.

[3] Cf. Friedrich Kambartel, *Erfahrung und Struktur, Bausteine zu einer Kritik des Empirismus und Formalismus*, Frankfurt a.M. 1968, p. 235.

[4] *Procli Diadochi in Primum Euclidis Elementorum Librum Commentarii, ex recognitione Godefredi Friedlein*, Leipzig 1873, reprinted Hildesheim 1967, 75, 6–11.

[5] Thoralf Skolem, "Some remarks on the foundation of set theory", *Proc. Int. Congr. of Mathematicians (Cambridge, Mass. 1950)* Providence, R.I. 1952, vol.2, p. 701f.

[6] Friedrich Kambartel, *Erfahrung und Struktur* (see Note 3), p. 241.

[7] Cf. Heinrich Scholz, "Die Sonderstellung der Logik-Kalküle im Bereich der elementaren logistischen Kalkülforschung", in: *Travaux du XIe Congrès International de Philosophie (Congrès Descartes), 6: Logique et Mathématiques*, Paris 1937, 20, pp. 40–42.

[8] The part of the text which follows was not contained in the original version of the lecture.

[9] Counting by means of the allocation of predicators and the special use of material and mass predicators are ways learned with one another. The frequent recourse by grammarians to the practice of counting in the characterization of the continuatives and collectives should not be regarded as constituting methodological circularity. On this problem see Willard Van Orman Quine, *Word and Object*, Cambridge, Mass. 1960, p. 95ff, and Kuno Lorenz, *Elemente der Sprachkritik*, Frankfurt a.M. 1970, pp. 209ff. and 228.

[10] On abstraction procedures see Paul Lorenzen, "Equality and Abstraction", *Ratio* **4** (1962), pp. 77–81, and my contribution "Gottlob Frege: Die Abstraktion", in Josef Speck (ed.), *Grundprobleme der großen Philosophen. Philosophie der Gegenwart* I, Göttingen 1972, [3]1985, pp. 9–46.

[11] Paul Bernays, "Die Philosophie der Mathematik und die Hilbertsche Beweistheorie", *Blätter für Deutsche Philosophie* **4** (1930/31), pp. 326–367, quotation on p. 351.

PAUL LORENZEN

GEOMETRY AS THE MEASURE-THEORETIC A PRIORI OF PHYSICS

The modern physicist knows that geometry goes back to Greek science. Thales (ca. 600 B.C.) is regarded as being the first geometer and Euclid wrote the classic geometry textbook about 300 B.C. Modern times have brought us Descartes' (1596–1650) analytical geometry, in which coordinate systems are used to represent geometric problems arithmetically. In the nineteenth century analytic geometry was further developed by Gauss and Riemann and by Felix Klein in the Erlangen Program (1872). Euclid's synthetic method was subsequently brought to the level of modern mathematics by Pasch (1882) and Hilbert (1899). This method no longer plays a role in modern physics, for the analytical method is adequate for all physical theories.

It is only since the emergence of relativity theory and the associated revolution in concepts of space and time (geometry and kinematics) that philosophers have been interested in clarifying the status of geometry as a theory *prior* to other physical theories. This *priority* is formulated using expressions like "a priori theories" or "protophysical theories". In order to clarify how geometrical theory is prior to physical theory, one must investigate synthetic geometry: investigating analytic geometry would only make sense on the basis of a yet-to-be-proved isomorphism between the domains of geometry and arithmetic.

Pasch and Hilbert regarded Euclidean geometry with the eye of the modern structuralist mathematician: that is, as though Euclid, by using a few simple propositions to *describe* the *structure* of the domain of physical objects (points, lines, and planes in space), had wanted to be able to prove all remaining propositions describing this structure on the basis of these few "axioms". Pasch and Hilbert assumed that the basic relations between geometrical objects are incidence, order and congruence. Hilbert selected fourteen axioms for plane geometry (ignoring here the axiom of completeness), carried out the construction of his geometry and the proof that it was isomorphic with analytical geometry, and showed at the same time that each of the fourteen axioms was indispensible. If any one of the fourteen axioms were to be omitted, then a pseudo-geometry would result (a non-Euclidean geometry as it is sometimes called).

Robert E. Butts and James Robert Brown (eds.), Constructivism and Science, 127–144.

One has only to read Euclid's first book (or at most the definitions in XI as well) to realize that Euclid's intentions in writing his textbook were different from those of the modern structuralist mathematicians from Hilbert to the Bourbaki group. Euclid constructed the forms of figures with a straight-edge and compass, that is, from straight lines (segments, rays, and lines) and circles. As the first three theorems of Book I show, he defined congruence using the straight line and circle as *basic forms* (*Grundformen*): two segments originating at a point M are congruent when their end points lie on the circumference of a circle with center M. If the segments do not originate from a common point then a construction must be carried out.

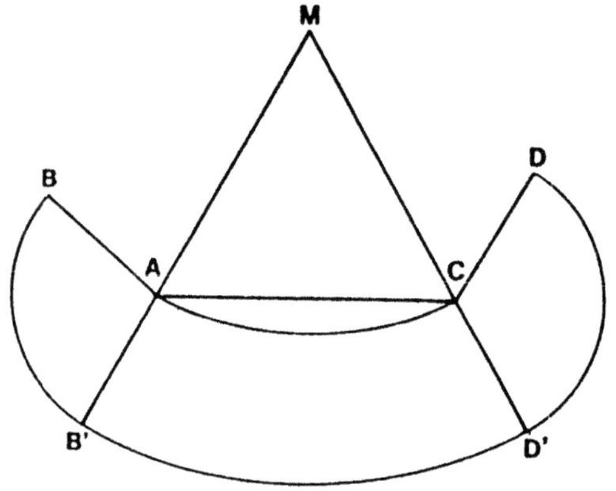

AB is congruent with CD just in case for the equilateral triangle AMC, B,B′ lie on the circumference of a circle with center A, and D,D′ lie on the circumference with center C, and finally B′ and D′ lie on a circle with center M. Euclid's very next theorem exhibits the flaw that this definition of congruence is not used. Consequently, this theorem (I,4) is an axiom. It is the theorem about the congruence of two triangles which correspond in two sides and the included angle. After I.4 Euclid never again makes a proof-relevant use of the term "to make coincident" (εφαπμοξειν). In Euclid's geometry (with the exception of theorems I.4 and I.8) nothing is said of motions (the transportation of rigid bodies); the forms of figures which may be constructed from lines and circles are discussed.

Since Helmholtz, however, physicists have usually conceived of geometry as a theory of the "free mobility" of rigid bodies (especially measuring sticks). This view also led to the Hilbertian axiomatization using congruence (and hence the "carrying over" (*Abtragen*) of segments and angles) as basic relations. Now we face the task of proving once more, with Hilbertian rigor, that geometry is a theory of constructible forms and figures.

This need not mean that, like Euclid, we must take straight lines and circles as basic forms. A simpler construction may be obtained if the *circle* is replaced by the *right angle*. Quite apart from systematic reasons (cf. the account of *protogeometry* in my book *Theorie der technischen und politischen Vernunft*, Stuttgart 1978) the origins of geometry in Thales' works speak in favor of this substitution.

In addition to the measuring rods and rope measures which had already been used to measure length in Mesopotamia and Egypt, the Greeks used, for the first time as far as we know, measuring discs to measure angles. These are our 360° protractors, most of which are still divided according to the ancient system into six parts whose further subdivision is sexagesimal. Since the French revolution, the protractor has been divided into four right angles R with a decimal subdivision, for example in 100 cR. A measuring disc serves to measure angles because the edge of the disc is a measuring strip divided into equal parts. Independently of the size of the disc and of the units on its edge, the *ratio* of the size of an angle to that of a right angle serves as the measurement of angle. Angle measurement has an absolute angle (R); length measurement on the other hand has no absolute length.

The "basic Thaletic figure" (*thaletische Grundfigur*) as Oskar Becker called it, is based on two straight lines intersecting at a point C

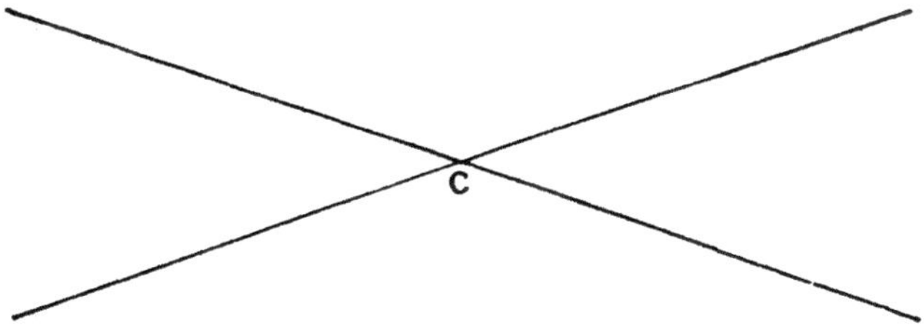

The angles formed by these lines are bisected using a protractor.

Since the angles to be bisected add up to 2 R, the bisectors of the angles are perpendicular to one another.

Taking a unit circle with center C and its four points of intersection

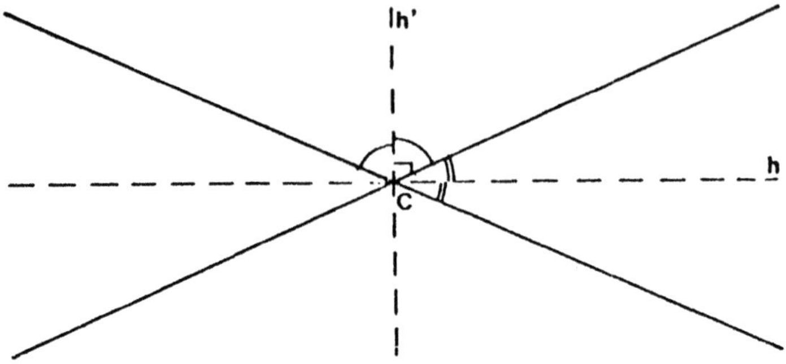

with the two intersecting lines, one obtains from any one such point its neighboring point of intersection by constructing perpendicular to the angle bisector a segment whose end point is the original point of intersection.

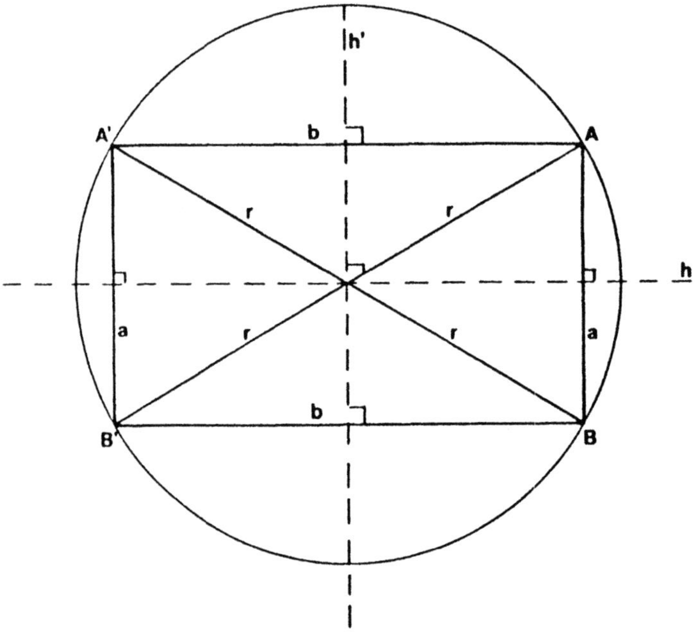

The *Thales quadrilateral* is a quadrilateral with the right angles shown in the diagram. The associated circle is a *Thales circle.* According to *Thales' theorem*, every Thales quadrilateral is a rectangle. The problem of geometry as a "theory" consists in proving such general propositions. Especially important for the justification of analytical geometry is proving the Pythagorean theorem: if c = 2 r then $c^2 = a^2 + b^2$. Proving this theorem was also Euclid's goal; it is the final theorem he proved in Book I. The basic Thaletic figure suggests defining the "congruity" of segments as either "side-equality" (*Seitengleichheit*) or "diagonal-equality" (*diagonalengleichheit*). Two segments CA and CB originating from point C are diagonal-equal just in case it is possible to construct from the free end-points a Thales quadrilateral on the extended segments.

Any two segments are "congruent" if they are each diagonal-equal to two side-equal segments. In order to distinguish it from protogeometrical congruity, this type of congruity is also termed "size-equality" (*Größengleichheit*).

On two rays originating at a point C, for any point A on one of the rays a point B can be constructed on the other ray, if one "carries over" (*abträgt*) the segment CA from one ray to the other (with a compass or even with a standard measure).

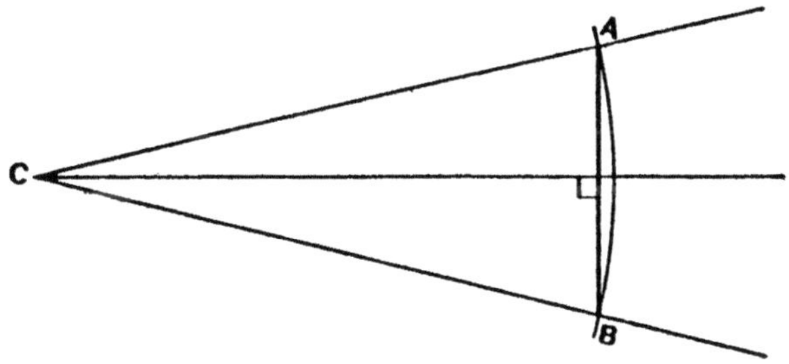

The angle bisector can also be constructed this way (as the perpendicular to AB through C). If one proceeds the other way around, beginning with the constructibility of the angle bisectors (which are defined as the two lines through the middle point of a Thales quadrilateral which are perpendicular to each other and to the sides of the quadrilateral), then the diagonal-equality is definable in a similar fashion.

These considerations lead to the construction of a "Thaletic" geometry, as I would like to call it, which constructs the forms of figures from the straight line and the right angle. For the constructibility of straight lines linking two points (*Verbindungsgerade*), perpendiculars *and* angle bisectors must be set up as axioms. Two points suffice as the basis of all constructions. Protogeometrically, these designate the end-points of a calibration measure (*Eichmass*), which is something like an "original measure" (*Urmeter*).

The technique of measuring length is reproducible only because it is invariant with regard to scale. Pairs of points which serve as bases of constructions cannot be geometrically distinguished from one another on the basis of the figures which may be constructed from them. These figures are even identical in form ("*similar*"). The implicit ordering of points on lines found in Euclid is unproblematic for Thaletic geometry. The ordering is justifiable protogeometrically and is invariant with respect to size. For this reason the usual ordering theorems, for instance, the two Hilbertian axioms can be taken as ordering axiom for Thaletic geometry as well:

- Exactly one of three points on a straight line lies between the other two.
- a line which intersects a side of a triangle but not a corner, intersects another side as well (Pasch's axioms).

There are two sorts of things to be described:

- points, A, B, C
- straight lines (lines, rays, segments) a, b, c ...

As basic relations we have incidence, $A \mid a$, order ABC, and perpendicularity $a \perp b$ (instead of ABC we may also write CBA, instead of $a \perp b$ also $b \perp a$).

Using the ordering axioms, both the segment AB and the two rays originating from a point A on a straight line a may be defined in the usual way. We call these rays $A \rightarrow a$ and $A \leftarrow a$. A segment AB is a *fully-oriented segment* if one of the points, for instance A, is marked and if one of the two rays, $A \rightarrow a$ and $A \leftarrow a$, which lie on the perpendicular to a through A, is also marked.

The "existence" of two points suffices as the basic axiom (1); (2) and (3) are the ordering axioms. Next come three construction axioms:

(4) For any two points A and B, the conjoining straight line (*Verbindungsgerade*) may be uniquely constructed: $A \vee B$.

(5) For any point A and any line a, the perpendicular to a through A may be uniquely constructed: $A \dashv a$.

(6) For any two rays, $C \to a$ and $C \to b$, originating at a point C, the angle bisector may be uniquely constructed: $(C \to a) \times (C \to b)$.

The final axiom is the *principle of form*:

(7) Figures which are constructed in the same manner from two fully oriented segments are geometrically (that is, in terms of the basic relations) indistinguishable.

In what follows I set out how we arrive at a proof of the Phythagorean theorem using the seven axioms of Thaletic geometry. Of Hilbert's fourteen axioms we employ only two ordering axioms. (The other two ordering axioms are provable on the basis of the construction axioms.) Ten axioms remain to be proved. A bit of scrutiny reveals that seven of these ten are trivial in our Thaletic geometry. The basis axiom and the constructibility of connecting line-segments deliver the three incidence axioms, and the four congruity axioms follow from the defining of size-equality rather than congruence as a transitive hull (*Hülle*) for side-equality and diagonal-equality. Only the congruity axiom for triangles (Euclid I.4) and the famous parallel axiom remain.

We have to show that the first two axioms mentioned above are – essentially thanks to the principle of form – provable in Thaletic geometry. Then there remains from the fourteen Hilbertian axioms only the Archimedian axiom: this is trivially true since, according to the construction axioms, analytically speaking, only numbers from the Pythagorean number field (*Zahlkörper*) (the set which for ξ always contains $\sqrt{(1 + \xi^2)}$) are constructible.

The parallel axiom turns out to be the following *parallel theorem*:

Two lines a, b have a point of intersection just in case for some line c $a \perp c$ but not $b \perp c$ is true.

Proof: If for a straight line c, $a \perp c$ and $b \perp c$ are true, then a and b do not intersect.

If in this construction the two rays $A \to a$ and $B \to b$ intersected (over the fully oriented basis AB) then – according to the principle of form – the rays $A \leftarrow a$ and $B \leftarrow b$ would intersect over the basis AB with the other orientation. This would contradict the uniqueness (*Eindeutigkeit*) of connecting line segments.

For the other direction, construct over AB a perpendicular a and a

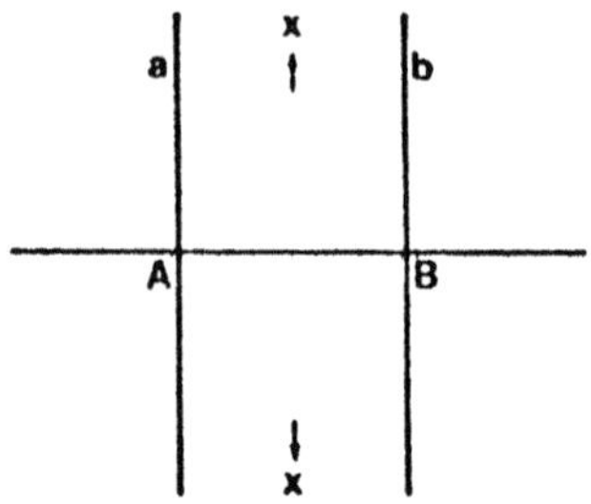

line *b* with an angle *β* smaller than *R*. (Such an angle is constructed by successively constructing angle bisectors of the angle *R* and the bisected angles. On *b* at least one point *C* may be constructed (for example, by taking the perpendicular to *b* through *A*). The perpendicular $C \dashv (A \vee B)$ would intersect *AB* in A′. Above *A′B* a right triangle with the hypotenuse angle *β* may be constructed. According to the principle of form, this is also true of the segment *AB*.

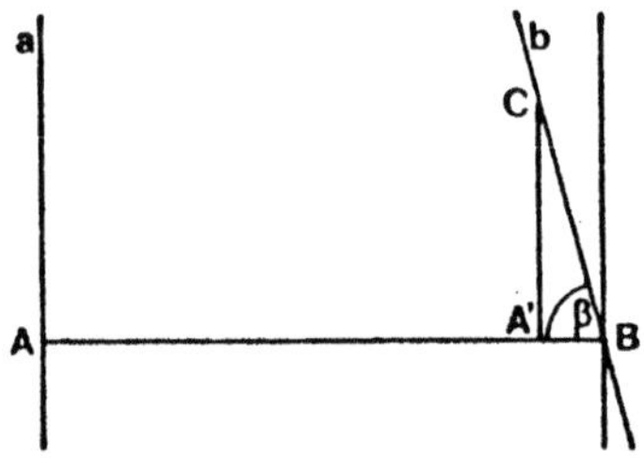

The parallel theorem permits us to prove the theorem of four right angles without using congruity theorems: If there is a line which is perpendicular to both of two other lines, then every line which is perpendicular to one of the two lines is perpendicular to both of them. Thales theorem can be derived from this result. This permits us to construct the square (with perpendicular diagonals) above every constructed segment.

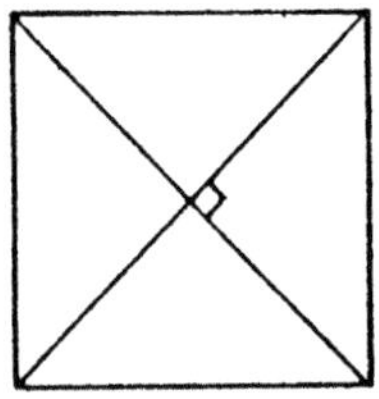

Next, by arbitrarily iterating, we get a grid:

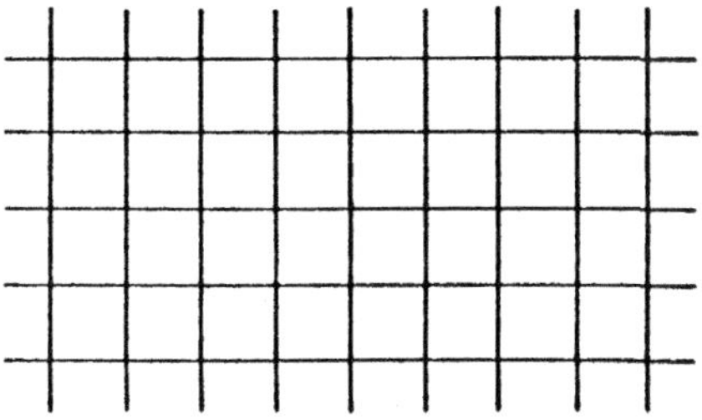

By bisecting the sides of the squares, the grid can be made arbitrarily fine. In a well-known manner, the length of the segments on any line in the grid can be defined relative to the basis unit (*Gitterbasis*) of the grid. In other words, Cartesian coordinates may be constructed for any (already-constructed) point. If another basic unit is chosen (for instance, so that the previous unit point (1,0) receives the new coordinates (λ, 0)), then the point whose old coordinates were (ξ, η) receives the new coordinates ($\lambda\xi$, $\lambda\eta$). The *tangent theorem* follows from this result:

If two right triangles have the same angle between the base and the hypotenuse (hypotenuse angle) then their short sides have the same ratio to their bases.

Proof. One constructs a grid so that one side of a right triangle lies on the horizontal axis ($\eta = 0$) and the intersection of the base and the hypotenuse is the origin (0, 0).

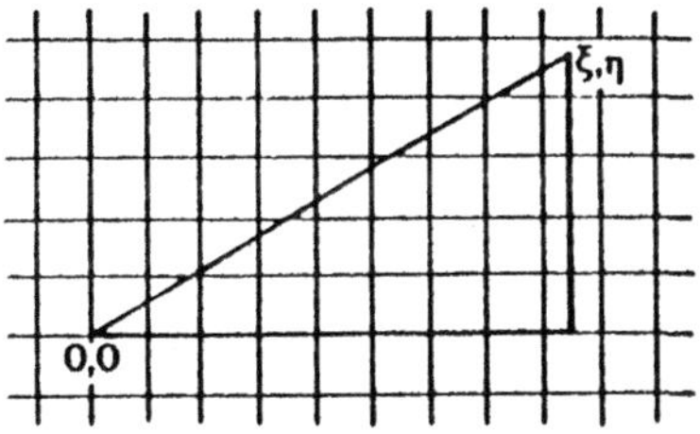

The hypotenuse thus has the end point (ξ, η). On some grid with different units this point has the coordinates ($\lambda\eta$, $\lambda\eta$). Thus there is some right triangle whose hypotenuse intersects its base at the origin

(0, 0) whose hypotenuse has the endpoint $(\lambda\xi, \lambda\eta)$ in relation to a grid with different units. According to the principle of form there is such a triangle relative to the original grid, that is, the point $(\lambda\xi, \lambda\eta)$ lies on the extension of the hypotenuse. The ratio of the sides is constant $\frac{\xi}{\eta} = \frac{\lambda\xi}{\lambda\eta}$.

Given the tangent theorem the Pythagorean theorem can be proved in four steps (the proof using the Hilbertian congruity axioms would thus involve a detour). First one can demonstrate the equality of alternate angles on intersected parallel lines

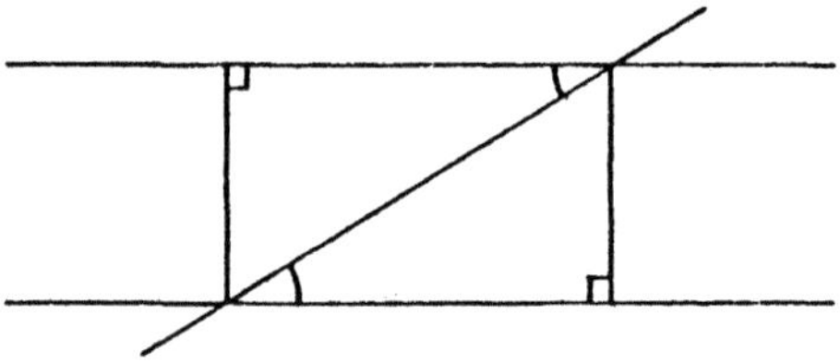

because the alternate angles belong to right triangles with equal sides. In particular, it follows that the sum of the hypotenuse-angles is R (90°). One can next prove the height theorem:

if a right triangle has height h and base segments (Höhenabschnitten) p and q, then $h^2 = p\,q$.

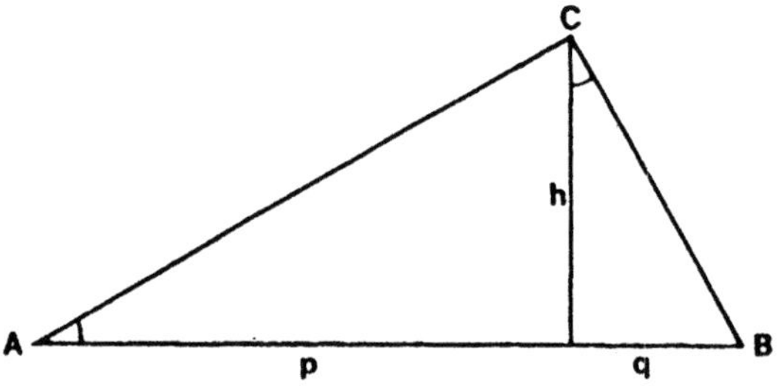

Proof. The angles marked at A and C are size-equal because they are both "hypotenuse angles", that is adding the angle at B to either results in a right angle. Thus $\frac{h}{p} = \frac{q}{h}$.

The Pythagorean theorem follows then for a right triangle with the sides a, b and hypotenuse c:

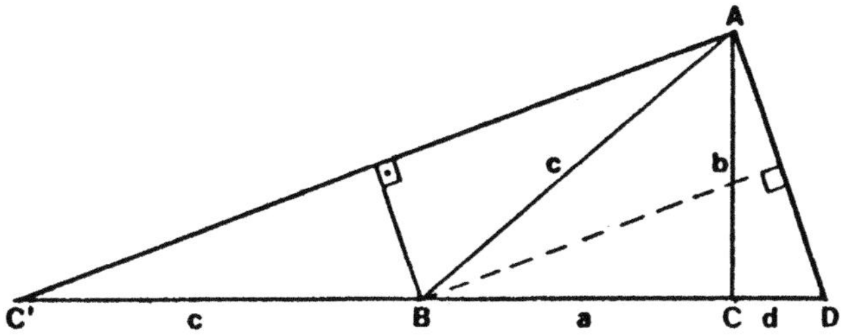

We extend side a by segment c through B to C' by constructing the angle bisector at B. The perpendicular through A to the second angle bisector [of B] intersects $B \vee C$ in C, resulting in CD, segment d. $C'AD$ is a right triangle according to Thales' theorem. Because $|BD| = c$, it follows that $d = c - a$. According to the height theorem $b^2 = (c + a)(c - a)$ Q.E.D

Using this theorem the coordinate transformations can be calculated for all grids, including those formed by rotations. At this point, we have reached the level of analytic geometry. This brings us to the isomorphism theorem:

For Thaletic geometry the domain of constructible elements is isomorphic with the domain of Cartesian coordinate geometry for the Pythagorean number field.

Expansion of the Pythagorean number field (for example, extending it to the Euclidean number field which contains $\sqrt{|\xi|}$ for all ξ) can be carried out by using constructive analysis. This problem, like the Hilbertian "completeness" problem, is an analytical, not a geometrical problem and is not addressed here. For physics, it is only important that Thaletic geometry sets out a theory of measuring length. Such a theory must enable us to check in reproducible ways whether or not the measuring sticks used in certain measurements are "rigid", that is, whether they deliver size-identical segments when transported. Thaletic geometry meets this criterion in that lines must only be checked for straightness and angles for perpendicularity. Using protogeometry's uniqueness theorems (*Eindeutigkeitssätze*) for planes, lines, and right angles, this checking can be shown to be a reproducible process. Such testing criteria as standard measures are not themselves checkable using length or other reproducible measurements. Therefore they are called measurement-theoretical *aprioris* of physics. "Geometry" is the name for the theory of length-measuring criteria, that is, for the definition of size-equality ("length-equality") using the size-independent basic forms of lines and right angles.

The systematic objection to the Euclidean basic form, the circle, is that one must be able to check whether the size of the circle changes after it is transported. The Thaletic definition of the size-equality of radii does exactly that. So much for geometry.

If we were agreed about the status of geometry as a theory of measuring length then presumably it would be simple to reach a consensus about kinematics, the mechanics of collisions, and dynamics from Newton to Einstein. Without a consensus about geometry, I can only somewhat dogmatically outline how one proceeds to dynamics if one has already accepted as one's goal understanding (*Beherrschung*) the movements of bodies.

I use the term "kinematics" as it was defined in the 1970 *Brockhaus Enzyklopädie*: "that part of mechanics in which the movements of bodies are investigated without attention to causal forces." In technology, kinematics is an important part of the theory of gear movement (*Getriebelehre*). A body has six degrees of freedom in mobility. Kinematic chains are fashioned from elementary pairs for which the mobility is reduced by at least one degree of freedom. A pair of planes has, for example, only 3 degrees of freedom; the pair consisting of a nut and bolt has only one degree of freedom; a chain with only one degree of freedom is called "constrained" (*zwangläufig*). (Compare W. Blaschke, *Ebene Kinematik*, München 1956.)

In order to calculate the velocities occurring in gearworks (*Getrieben*) one must define clocks with which scale invariant movement-durations can be measured. (For brevity's sake, I refer the reader to the definition in my *Theorie der technischen und politischen Vernunft*). If we limit ourselves to digital instruments, that is, to metronomes (*Taktgeber*) with digital meters, then such a device is a clock if, starting in any arbitrary position, it keeps the same rhythm as all copies [of it].

If G_1 and G_2 are two clocks of one type and G_3 and G_4 are two clocks of another type, then for positions α, β, γ, δ, G_1^α and G_2^β are synchronic, as are G_3^γ and G_4^δ. Therefore

$$\wedge \alpha, \beta, \gamma, \delta, G_1^\alpha : G_3^\gamma = G_2^\beta : G_4^\delta$$

In accordance with quantification and equality logic

$$\vee \beta, \delta \wedge \alpha, \gamma\ G_1^\alpha : G_3^\gamma = G_2^\beta : G_4^\delta$$

and

$$\wedge \alpha', \gamma' \wedge \alpha'', \gamma''\ G_1^{\alpha'} : G_3^{\gamma'} = G_1^{\alpha''} : G_3^{\gamma''}$$

The pace ratio of any two clocks are the same: we have a scale-invariant duration measure (or time measure).

Kinematics is the theory of forms of motion (*Bewegungsformen*) which are definable using such constructions. For historical and systematic reasons I suggest the mechanics of collisions as the transition step from kinematics (in which *mass* plays no role) to dynamics. Departing from the sort of physics which had been done up to that point in history, collision theories considered bodies apart from their weight. Abstracting from the weight of bodies happened on a practical level when billiards was developed in Italy in the 16th century. The laws of collision developed in the 17th century are the theoretical counterpart of the game. Billiards realized completely elastic collisions but John Wallis showed in 1670 how they could be reduced theoretically to completely inelastic ones. In a completely inelastic collision of two particles along a straight line there are two velocities, v_1 and v_2, before the collision, and a common velocity, v, after the collision. This collision process can be technically realized in such a way that in repeated collisions of two particles (that is, sufficiently small balls) the velocity v will prove to be a function of v_1 and v_2. The *law of collisions* is a functional term S with $v = S(v_1, v_2)$. In the field of classical-technical velocities of bodies, in ballistics for example, it can be shown empirically, by measuring, that S is a homogeneous function $S(\lambda v_1, \lambda v_2) = \lambda S(v_1, v_2)$ if $\lambda > 0$. The earth serves – approximately – as an inertial system.

Measurement (this is empirical physics, at the moment) enables us to define the ratio of two masses:

$$\frac{m_1}{m_2} \leftrightharpoons \lim_{v_1, v_2 \to 0} \frac{v_2 - v}{v - v_1}$$

Hence we can measure the mass ratio for sufficiently small velocities. The definition of the limit is scale-invariant. The classical *momentum theorem* follows from this definition:

$m_1 v_1 + m_2 v_2 = (m_1 + m_2)\, v$ for a completely inelastic collision. The quantity "mv" is called "momentum".

The step to dynamics is made by defining "force" in terms of the changing of momentum over time – that is, *as if* the body were being struck. This leads to the Newtonian program in vector notation:

$$m \frac{\mathrm{d}V}{\mathrm{d}t} = \Sigma_j K_j$$

Suitable force laws can be found which permit us to calculate observed or expected changes in momentum. In particular, the success of the law of gravitation $K \sim mM/r^2$ in astronomy assured the Newtonian program its central place in 18th and 19th century physics.

Difficulties did not emerge until the end of the 19th century when electrodynamics began to be introduced into the Newtonian program through Maxwell's equations. The theory of electricity began with the electrostatic definition of charge ratios. Charged bodies were produced in the laboratory and the force (change of momentum) exerted by these sources on charged "test bodies" was measured. It was empirically verified that in every position the ratio of forces K_1, K_2 depended on the two test bodies, and was independent of the charge of the source and of the source's distance from the test body.

Such measurements enabled Coulomb to define a charge ratio:

$$\frac{q_1}{q_2} \leftrightharpoons \frac{K_1}{K_2}$$

Taking the source as our starting point, one might say that it exerts a "field-strength" E on every point in space so that on any particle with charge q, a force $K = qE$ would be exerted.

The transition from talk of particles to talk of electric fields is made when for every position in space a charge intensity ρ is defined. A force intensity k is calculated according to the *Coulomb formula* $k = \rho E$.

For moving charges, the force intensity k is calculated according the *Laplace formula* $k = J \times *B$. Here, $*B$ is the so-called magnetic flux density (*Flußdichte*) and J is electric current density (that is, charge density $\times$ velocity). The combined form of the *Laplace* and *Coulomb* formula is the *Lorentz formula*: $k = \rho E + J \times *B$. The vectors E and B can be calculated using Maxwell's equations. B is called the "magnetic bivector" (*magnetischer Bivektor*) and is the orthocomplement (*Orthokomplement*) of $*B$. The value of E and B depend on charge and current density values.

In the notation of alternating differential forms, these are two homogeneous equations: $\mathrm{d}\ B = 0$, $\mathrm{d}\ E + \dot{B} = 0$. (The dot over B is Newton's notation for differentiation over time.) For $\delta = *\mathrm{d}*$ there are two proportions: $\delta B \sim J$ (for $\dot{E} = 0$) and $\delta E \sim \rho$. In addition, there is the charge constancy $\delta\ J + \dot{\rho} = 0$.

The following proportionality coefficients are defined: $\delta B = \mu_0 J$ (for $\dot{E} = 0$) and $\delta E = 1/\varepsilon\rho$. Everything takes place in a vacuum.

Because of the charge constancy we follow Maxwell in using the following formula when $\dot{E} \neq 0 : \delta B = \mu_0\ (J + \varepsilon_0 \dot{E})$.

The product $\mu_0\ \varepsilon_0$ somewhat surprisingly delivers the speed of light c using the definition $1/c^2 \leftrightharpoons \mu_0\varepsilon_0$. In this way, the inhomogeneous Maxwell equations become

$$\delta B - \frac{1}{c^2}\ \dot{E} = \mu_0 J$$

$$\frac{1}{c^2}\delta E = \mu_0\rho$$

When we want to do Newtonian mechanics, it is annoying that "velocity" occurs in these equations. If for our space and time measurements we use a cubic coordinate-system that we imagine to be fixed relative to our earth-bound laboratory, and if we neglect momentarily the rotation and revolution of the earth, then we have an inertial system, that is, the classical momentum theorem holds. A reference system moving uniformly relative to the earth must then also be an inertial system.

For Maxwell's equation, Lorentz and Poincaré recognized that these reference systems can only be coordinate-invariantly formulated when one introduces a new metric (in Riemann's sense) for the continuous manifolds of space-time points (x^1, x^2, x^3 and $x^4 = t$) belonging to the laboratory system. The fundamental tensor, G_{mn}, is defined such that in the laboratory system the following equation holds:

$$G_{mn} = \begin{pmatrix} 1 & & & \\ & 1 & & 0 \\ & & 1 & \\ 0 & & & -c^2 \end{pmatrix}$$

Instead of the space vectors we have used up until now, the following four vectors are introduced using $\mathbf{B} = B + E\,\mathrm{d}t$ and $\mathbf{J} = J - c^2\rho\,\mathrm{d}t$. We replace our old Maxwell's equations with the following system of equations in four variables: $\mathbf{dB} = 0$, $\delta\mathbf{B} = \mu_0\mathbf{J}$. Even this can be more briefly written, if a four-potential (*Viererpotential*) $\mathbf{A}$ is defined as $\mathrm{d}\mathbf{A} = \mathbf{B}$. There remains only one equation: $\delta\mathrm{d}\mathbf{A} = \mu_0\mathbf{J}$.

This can all be regarded as mere mathematical elegance. However, Einstein saw that these formulas suggest a revision of the classical momentum theorem. Instead of Lorentzian force-intensity, k, he used a four-force-density $\mathbf{k}$ defined as $\mathbf{k} = \mathbf{J} \times {}^*\mathbf{B}$. In Minkowski's notation one revises the Newtonian equation

$$m\,\frac{\mathrm{d}V}{\mathrm{d}t} = \Sigma_j K_j$$

and gets instead the Newton–Minkowskian equation for suitable four-force-terms:

$$m\,\frac{\mathrm{d}\mathbf{W}}{\mathrm{d}\tau} = \Sigma_j \mathbf{K}_j$$

Riemann's investigations of continuous manifolds with metrics defined by fundamental tensors result necessarily in the definition of a four-momentum $m\mathbf{W}$ and an "arc-time" (*Bogenzeit*) τ instead of the classical time coordinate, t. For the arc-length (*Bogenlänge*) σ, the following equation holds for every curve parameter, s.

$$\mathrm{d}\sigma = \sqrt{\left(G_{mn}\,\frac{\mathrm{d}x^m}{\mathrm{d}s}\,\frac{\mathrm{d}x^n}{\mathrm{d}s}\right)}\,\mathrm{d}s$$

The arc time, τ is defined as $ic\tau = \sigma$, and the four-vector $\mathbf{W}$ (four-momentum per mass) as

$$\mathbf{W}^m = \frac{\mathrm{d}x^m}{\mathrm{d}\tau}$$

In the space coordinates ($m = 1, 2, 3$) **W** is distinguished from the classical velocity vector

$$V^m = \frac{dx^m}{dt}$$

by the famous factor:

$$\beta_v = \frac{1}{\sqrt{\left(1 - \frac{v^2}{c^2}\right)}}$$

Because $\beta_0 = 1$, Newtonian dynamics is an approximation of Newton–Minkowskian dynamics.

It would be confusing to maintain that the kinematic velocity vector V was replaced by the four-velocity **W** in the course of revising kinematics. It is better to say that the classical momentum, mV was replaced by the four-vector $m\mathbf{W}$ in the course of revising collision mechanics. It is difficult to recognize the difference when one just calls collision mechanics "kinematics". Unfortunately, this is usually the case. We lack then a word for the a priori theory of measuring time and movement.

Einstein recognized early on (in 1907) that it was impossible to treat gravitation in a Newton–Minkowskian program using a suitable four-force. He had what he himself called "the best idea of his life" when he recognized that one did not need any force concept to account for weight in a homogeneous gravitational field. When the roofer falls from the roof, he is (from the physicist's viewpoint) in the fortunate position of finding himself in a gravity-free Newton–Minkowskian inertial system. All bodies are in a state of constant acceleration. It is quite strange to explain this by saying that on the one hand force = mass m times acceleration a; and on the other hand, gravitational force = mass m time gravitational constant g. For, when $m > 0$, $a = g$.

In homogeneous gravitational fields, one also needs no four-force (*Viererkraft*) for the Newton–Minkowskian program. Before moving into the Newton–Minkowskian program, one adopts a coordinate system which is accelerating at the rate of g relative to the laboratory. The physicist lets himself fall from the roof in his imagination!

The general theory of relativity solves the problem of determining coordination transformations leading to a Newton–Minkowskian intertial system at any position of the field. Einstein and Hilbert found the solution in 1915: A tensor field G_{mn} can be calculated from the current density tensor T_{mn} using the field equation $R_{mn} - \frac{1}{2}G_{mn} \cdot R \sim T_{mn}$. ($R_{mm}$ and R are in this way reductions of R^k_{mln}, a Riemann tensor which is determined by G_{mn} alone.)

Corresponding to the potential-equation of electrodynamics, G_{mn} is a "potential" of the desired transformation to a local Newton–Minkowskian inertial system.

S. Weinberg describes this interpretation of general relativity theory in his 1972 book *Gravitation and Cosmology*. Any further talk of revising geometry is but a misuse of the geometrical language.

PETER JANICH

THE CONCEPT OF MASS

In the analytic-empirical philosophy of science, technical terms not explicitly defined are referred to, within a two level model of scientific language, as "theoretical concepts" or "theoretical terms". After it had proved impossible to reduce physical terminology to a non-theoretical observational language by means of explicit definitions or complete interpretations of all the terms of a theory, the term "theoretical concept" in the work of C. G. Hempel was first interpreted to mean that such terms cannot be defined or determined explicitly and independently of the complex of the other terms in a particular theory. Nor can they be defined independently of the empirical validity or corroboration of this theory.[1]

This view of theoretical concepts underwent a radical change in the model-theoretic approach of J. D. Sneed. It is, however, accepted by adherents of the widely adopted analytic-empirical philosophy of science that physical terms, and particularly those which represent metric concepts, are theoretical, that is, they cannot be explicitly defined. This paper draws attention to certain previous assumptions which support the view that the concept of mass in physics is a theoretical concept. These presuppositions will make it clear that there are good reasons for presenting another view of the concept of mass. By employing the formal rigour of the theory of scientific language I hope to go beyond the analytic tradition and to achieve both methodological precision and a perspective which is closer to physics as it is practiced. In the first part of this paper, I shall list and criticize some of the reasons for assuming the theoretical status of the concept of mass in analytic philosophy of science in order to then draw attention to a problematic nature of basic presuppositions in the analytical approach. In the second part, I shall discuss generally acknowledged problems of defining mass from a different theoretical perspective and suggest a non-circular operative definition of mass.

I

The current interests of analytic philosophers of science are primarily

Robert E. Butts and James Robert Brown (eds.), Constructivism and Science, 145–162.

directed toward the approach of J. D. Sneed, in so far as these philosophers have not been disconcerted by P. K. Feyerabend.[2] This view of physical theories, which is achieved by means of model theory and is equivalent to the elimination of theoretical terms by means of an improved Ramsey Sentence Method, ultimately retains the assumption of the non-definability of all basic physical concepts, and hence the idea of theoretical terms.

It is therefore worth examining earlier views on the subject, as essentially developed by R. Carnap and C. G. Hempel. One can then see more clearly the reasons why physical terms are considered to be theoretical. The fundamental decision in favour of physics as an example of scientific rationality worthy of philosophical analysis which here becomes apparent is then repeated in specifically emphasized form in the work of Sneed. I shall first mention three of the numerous arguments for assuming only a partial interpretability of the vocabulary of the theoretical language in the last phase of the program of reduction of a theoretical language to an observational one. These, if they may be assumed to be true at all, are true for all metric concepts:

(1) Metric concepts exhibit idealization, since, in physical theories, quantities are expressed in real numbers although the results of measurement can only be expressed in rational numbers.[3]

(2) The formation of metric concepts involves hypothetical generalizations of experimental findings. This is the case, for example, when the transitivity of the equality of measuring results is assumed for measured quantities.[4]

(3) A particular metric concept can be formed by various operational definitions. In his critique of Bridgman, for example, Hempel mentions the measurement of electric resistance employing the Wheatstone Bridge and Ohm's Law respectively.

This list of arguments for the theoretical status of all metric concepts does not claim to be exhaustive. These reasons are, however, also particularly valid for every metric concept of mass.

The following are concerned solely with the theoretical character of the concept of mass:

(4) The well known problems of the definition of mass within the framework of classical mechanics. We assume that means of measuring both length and time are available. Newton's circular definition of

mass is the best known, namely, that mass is the product of density and volume. On the other hand, familiar definitions from textbooks on physics presuppose the inertial system of reference. These definitions merely conceal, with varying degrees of skill, the circularity of the argument when they state that inertial systems are defined by freedom from forces, whereby a definition of force in fact presupposes a measurement of mass.[5] These problems, which have been recognised for at least two hundred years, led investigators even in the last century to the view that only the system of mechanical principles as a whole, if possible in the axiomatic form, could be meaningful and valid.[6] All that was then necessary to reach the view that actual measurements of mass were models of a formal mechanics was a vocabulary originating from metamathematics and logic as well as adherence to the principles of formalistic mathematics.

Before I embark on a criticism of the above four arguments for the theoretical status of the concept of mass, let me first make a preliminary comment on the programmatic division of the language of physics into a theoretical and an observational language.

It is practically impossible today to find a proponent of analytic philosophy of science who believes in the possibility of a completely theory-independent observational language. Whatever the individual objections to the concept of a pure observational language are they all emphasize that even the simplest observations in physics are described in a language that is 'loaded with theory'. This criticism of the old two-level model on the language of science is, on the one hand, not sufficiently extensive and on the other hand goes too far. It is deficient in that it only examines the 'loadedness' of the observational language using terms that originate from *physical* or *other empirical* theories. It goes too far in that it suggests that there can be no observational language free of empirical terms. In my view, the objection to the two stage model of the language of science should, in its correct form, state that a division of the language of physics into two parts is too narrow for the following reason. Only those observations which are made with instruments play an essential role in physics. Thus, the properties of instruments are always constitutive for observation results formulated in a particular language, that is, those properties that are artificially and intentionally planned and constructed and which must be held constant during an experiment or measurement process. This applies whether we record pointer-positions from a scale or whether, in the case of instruments like telescopes, microscopes or interferometers, we obtain non-numeric results. The physicist-observer is a competent observer only when he is sure of certain properties of his instruments or when he is in position to assure himself of them. A complete observational proposition, therefore, always takes the form: the state of affairs S is indicated by an instrument with the properties $P_1, \ldots, P_n$ under the conditions $C_1, \ldots, C_m$ (by conditions $C_1, \ldots, C_m$ are to be understood the relevant physical parameters of an experiment, which must be determined with other instruments). In part the predicators designating properties $P_1, \ldots, P_n$ of an instrument are in every case, neither terms belonging to an empirical physical theory nor part of the observational language in the sense that they refer to naively perceptible, naturally given qualities independent of description in scientific terminology. They rather denote properties of instruments linked with the aims of their users. The competence of the user of an instrument consists in his being able to explicitly formulate the aims which he is pursuing in using the instrument. If the division of the scientific language into groups such as observational terms and theoretical terms is to be at all meaningful, we require at least one further group of terms designating properties of instruments.

Thus we have provided a basis for our criticism of the three arguments for the theoretical status of metric concepts outlined above.

Ad (1) The logical gap between measuring results expressed in rational numbers and those in computational expressions given in real numbers in theories is the central issue in the old problem of the application of mathematics to the objects of experience and since the discovery of irrational numbers one that is probably unbridgeable within the framework of the empirical program of reconstruction of metric concepts. The 'theoretical' objects, e.g. points, distances and angles, in the sphere of the measurement of length, cannot be obtained by logical abstraction via operations with measuring instruments marked in standard units. It may be supposed that any analytical philosopher of science would concede this. There are still those, however, who adhere to the empirical point of view originating with H. v. Helmholtz, according to which, measurement can be reduced to the counting of standard units.[7] This is all the more surprising since the reproduction of units to be counted undoubtedly presupposes a great deal of empirical scientific knowledge based on measurement.

The belief, however, that the approach to the metrization of physical quantity begins with the counting of standard units, oscillation of a pendulum or standard weights is neither a natural law nor analytically justified. It is, on the other hand, possible to formulate instructions for the reproduction of *forms* such as straight edges on measuring sticks, uniform motions of pointers on clocks and, as will be shown below, of the material form of homogeneous density, without recourse to the results of measurement. This is a decisive advantage compared with instructions for the reproduction of standard units for the measurement of time, length and other quantities. Within the framework of *protophysical* approaches, the problem of idealization – the transition from reference to real bodies to mathematical theory – may be regarded as solved. This solution also adequately covers the operative definition of equivalence relations as well as procedures for the determination of ratios of quantities.[8] Thus an idealization problem only arises as an argument for the theoretical status of metric concepts, if one adheres, without good reason and against one's better knowledge, to the empirical program that the definition of measuring quantities begins with the counting of units of measurement.

Ad (2) Operative definitions of measuring quantities allegedly contain hypothetical generalizations, for example, the view that the transitivity of equality for all measuring quantities can only be known by empirical testing. It is, however, possible to agree with the early empiricist v. Helmholtz and to argue against the view of the more recent empiricists Hempel, Carnap and others that the universal propositions in question are not hypotheses to be empirically tested, but are determined by the correct functioning of instruments. Hence, in non-empirical terms, they should be placed as *norms* at the beginning of a theory. They should not be referred to as hypotheses, since it is, generally, only possible to test them empirically if they are already valid. In the case of the transitivity of weight-equality on symmetrical scales, one requires either a set of weights gauged with the aid of scales, which are already correctly functioning and whose transitivity has already been checked, or one must presuppose that the scales are already functioning, in the sense of the transitivity of weight-equality, in order to be able to test the allegedly empirical proposition about the transitivity of weight-equality.[9] It is, therefore, erroneous to draw conclusions about the empirical status of measuring quantities from the construction of their logical properties with the help of universal quantifiers. Logical properties of measuring quantities are rather determined by the norms prescribing the functions of instruments, for an understanding of which, a special theoretical language is as unnecessary as it is in the case of the norms of the highway code.

Ad (3) The possibility of interpreting a particular "theoretical" term by means of various measurement procedures only arises when it is overlooked that measurement procedures are presupposed when a theory is even tentatively formulated. In principle, however, a single measurement procedure is sufficient for its formulation. The equivalence of measuring results from various measurement procedures can of course depend on empirical theories, as is clearly illustrated by means of the various procedures employed in the measurement of electrical voltage. But the problem which arises here, is that there must, to a certain extent, be a superordinate theoretical concept, quasi 'over and above' the various realizations as if for a natural property. This is but the result of the analytical procedure to analyze physics merely in the form of its theories. This aspect will be more closely criticized in what follows.

Ad (4) This case differs from the above three. It may be legitimately argued that so far there have been either no explicit definitions of mass which contain non-theoretical components, or there are only such definitions – such as those of B. Thüring[10] and P. Lorenzen[11] which contain stronger presuppositions in that they contain proper names in the logical sense (for the earth or an astronomical fundamental coordinate system).

By means of this argument put forward especially as a reason for the theoretical status of the concept of mass, it is possible to demonstrate even more clearly than in the case of the other three arguments where the responsibility for the problem of "theoretical concepts" lies. If there is agreement that, at least up to now, we have no explicit definition of mass independent of a theory of mechanics whose validity is presupposed, a philosopher of science can react in one of two ways. He can regard a part of physical theory as being incomplete and then, with the special competence of the philosopher of science, embark on detailed theoretical work in order to help find a solution to the problem of definition which has arisen. Alternatively, he can stick to the dizzy heights of meta-science, making no claims to practical relevance, invent complex descriptions of the relationship between theory and laboratory practice and hence by means of artistic explanations retrospectively confirm the validity of the claim of physics to scientific status. The analytical philosophy of science has opted for this second solution. Here it is not the case that alternative formulations of physical theories are sought in order to solve the problems of definition, but rather that one seeks appropriate descriptions of what can be read in physics textbooks. Although, in this case, the philosopher of science decides in what sense physics may claim to be scientific and thus makes science dependent on his metatheoretical artistry, his approach is occasionally claimed to be the more modest one and the rightful domain of the philosophy of science. By choosing this path the philosopher of science is said to avoid interfering in the specialist sciences. But the question may remain open as to whether these objections, which are occasionally raised with regard to protophysics, are justified. In particular, it may remain open whether it is more modest, after an analysis of the capacity of modern physics, to supply methodological instruments in order to validate its claim to being a science, or, whether after an analysis confining itself exclusively to the theories of physics, to first formulate its de facto validated and hence a fortiori demonstrable claim to scientific status.

Compared with the earlier approaches of Carnap and Hempel, the basic assumption of analytical philosophy of science as proposed by Sneed, takes a different and more radical form. Whereas Carnap and Hempel took physical theories in general as their starting point, Sneed relies on specific formulations of physical theories as found in textbooks. Hence, at best, he supplies us with a restrospective confirmation of physical theories by means of artistic explanations in a much wider sense. Following W. Stegmüller, who calls Hempel's theoretical concepts "theoretical in the weak sense", and Sneed's theoretical terms "theoretical in the strong sense",[12] I should like to term Hempel's view, and also that of Carnap, "affirmative in the weak sense" and Sneed's "affirmative in the strong sense". For reasons of brevity I would also suggest another term. Stegmüller's combination of Sneed's approach with T. S. Kuhn's understanding of the history of science will here be referred to as 'superaffirmative'. Hence the fundamental assumption of analytical philosophy of science, which gives rise to theoretical concepts, may be described as follows. The weakly affirmative view assumes that rationality may be found in the contents of physics textbooks. The strongly affirmative view presupposes that rationality may be found in the particular formulations of physical theories. The superaffirmative view ultimately starts from the rationality of the entire history of physics and its results. This last view, held by Stegmüller, has its origins in a decision for aims which are neither discussed nor confirmed. 'The harmony between metatheory and experience (is) to be restored in such a way that the model of the rational behaviour of the scientist is replaced by a more adequate concept of rationality.'[13]

In all three cases of the affirmative basic assumption, we must question assumption over supposition: for the formulation of concepts of rationality or scientificity the only remaining task is to demonstrate their adequacy with regard to the content of physical theories, to their individual formulation or to the entire development of the history of physics. This is the same as assuming that rationality may generally be presupposed in the case of physics. Even if it is not denied that physics still provides the best example of an experimental science, this assumption remains questionable. The theories of physics and above all, the history of physics (in the sense of the development, not the description) is the work of human beings and thus, the assumption that rationality and only rationality have succeeded in physics is just as dubious as the opposite assumption. The alternative to such prejudices is to assume that physics too has developed as a mixture of the rational and the irrational. In this case, it is the task of both physicists and philosophers to agree as to what rational is supposed to mean and, as a second step, to examine physics as it stands to see what is rational about it. The tolerance of definitional deficiencies with regard to the concept of mass in both classical and nonclassical physics cannot, however, be counted among the rational achievements of physics and the history of the philosophy of science.

II

The well-known problems of definition concerning inert mass are given extensive treatment in the literature. These problems first arose in classical mechanics but have not been fundamentally solved even in modern theories. The result is that it has become fashionable to assume, without good reason, that they cannot be solved. Such problems are essentially attributable to the fact that all definitions supplied hitherto contain dynamically formulated isolation conditions for those bodies or for that frame of reference, for or within which 'mass' is defined. Isolation is, for example, defined as free mobility or freedom from forces on the part of the body in question or as the 'neutrality' of experimental bodies with respect to all known fields. Such isolation conditions are intended to define an inertial reference system. If it is wished to obtain an operational realization of force free movements, it is hoped to obtain this dynamic 'isolation' by compensating for those forces from which the defining system cannot be isolated. However, this isolation can itself only be defined and realized if forces or mass can already be measured or technically mastered.

This argument, which expresses the circularity of existing definitions of mass by freely moving bodies is, within certain theories of definition, complemented by a proof of non-definability based on specific formulations of particular physical theories. This proof has already been employed by the analytical philosophy of science as an argument against an operative definition of inert mass.[14] However, those proofs can with the aid of Padoa's proposition, for example, be employed only in the context of already existing theories. They refer here to a specific vocabulary in the form of an axiomatic physical theory, for example. They are only concerned with theory-immanent problems of definition and they do not represent a general proof of impossibility of an operative definition free of formalistic limitations and independent of presupposed axiomatic theories. As a glance at such proofs within theory of definition shows,[15] they are no more a proof of impossibility than the non-derivability of a proposition in one particular theory would imply the non-derivability of that proposition in any theory.

I see a genuine alternative to a definition of mass which is unaffected by existing proofs in theories of definition in dispensing with the notion of the free movement of bodies (which is responsible for the problems of definition described above). In order to justify this step and the resulting definitions I should like to borrow from the history of science. The leap from Aristotelian dynamics, where forces were viewed as causing velocity, to the inertial-mechanics of the seventeenth century where forces were seen as causing changes in velocity is clearly expressed in one of Galileo's writings.[16] After falling movements in an oblique plane had been recognised as involving acceleration (in the sense of a change of speed proportional to time) and climbing movements as their converse, Galileo regarded horizontal movement at a constant speed as a borderline case between falling and climbing. Reductions in velocity due to friction are treated in the same manner as in the case of falling and climbing movements, that is to say, the less the friction, the better the phenomena of accelerated or constant movement are demonstrated. Independently of the evident fact that 'horizontal' is only definable with reference to the *earth*, Galileo is concerned with the accelerating or braking effect of gravitation in falling and climbing movements. Hence, it follows that, for horizontal movements, the earth, of course, remains the system of reference. Even from a modern point of view, no definitional problems arise here.

The transition to the so-called classical principle of inertia in the work of Huygens and Newton, for which the question of the system of reference can no longer be answered, was fatal from a theoretical definitional point of view. Borrowing from the history of science, I thus return to those unproblematic movements describable within the system of reference provided by the earth. It will here become clear that an attempt of define mass can be developed to such a degree that, by statement of additional measures, the earth as a system of reference is rendered superfluous in a definition of mass.

The following definition is intended to be both operative in the strict sense, and non-circular, that is to say, it should consist of a catalogue of instructions actually fulfillable under existing conditions. Thus, it may not assume any technical measures which are only possible on the basis of the successful measurement of mass. Anticipating a physical terminology which must first be reconstructed, this means that the following definition can only be applicable given the known effects of gravity, a medium like air or water and with expanded bodies, not necessarily homogeneously dense ones, for example.

As a methodological maxim it should be borne in mind that, in the logical sense, this approach is free from proper names in that the proper name 'earth' does not occur in the definition of mass. Only then can we hope, in addition to the systematic tasks of defining mass, to suggest a method for the reconstruction of classical mechanics.

I here assume an operatively justified geometry and a time-independent kinematics, that is to say, a purely geometrical comparison procedure for simultaneous movements. Both assumptions are unproblematic in the light of existing protophysical theories for the measurement of length and time.[17] A real object, that is, an object not merely conceived of as a point, may be considered to be *kinematically guided* if its path (direction of movement) is determined and the body is kept free from rotation by means of a rail, for example. A body may be spoken of as being *dynamically guided* if it is kinematically guided, and if its velocity, (its respective speed at any moment) is determined by 'traction', for instance by traction via a rope. The expression 'traction' is here operationally defined with the example of the human action of pulling a body with a rope.

We have here reached a point where the definition procedure favoured by analytical philosophy of science may be abandoned and an operative start on the construction of the terminology in the strict sense may be made. The truth of a statement to the effect that a body is subjected to 'traction' because it is pulled by a human being is here not tested but produced. Thus, in case of doubt, every competent speaker can ascertain if pulling is taking place through his own actions or, alternatively, can ascertain this from the fact that the rope is taut. In order to make unambiguous statements of this sort the explicit establishment of any system of reference is irrelevant. Were one, at this stage, to have at one's disposal a defined way of talking about forces then one might say that it is a sufficient definitional condition for a force to act on the body via a taut rope. The following symmetrical arrangement will be called a 'rope-balance'.

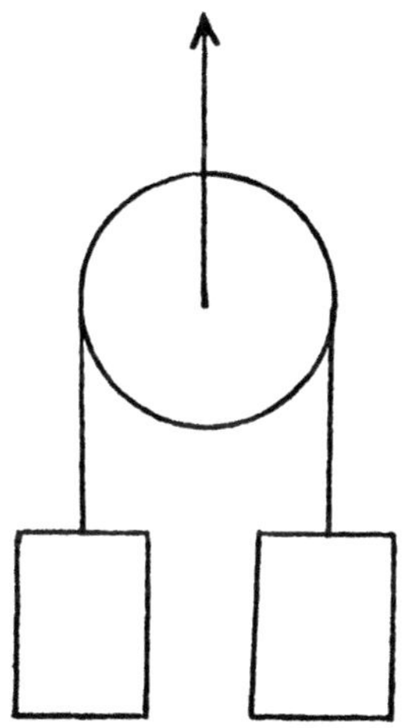

It is possible, by geometrical means alone, to ascertain the rigidity of the rope in the sense of the constancy of its length as well as to determine the symmetry of the balance and the parallelness of all of the traction ropes. For purposes of terminological simplification we shall now restrict talk of a 'rope-balance' to those cases where traction is actually taking place. Two bodies K1 und K2 may be referred to as being *tractionally equal* if the wheel of the balance is symmetrical and at rest with regard to the direction of traction of the ropes. In this case a functional norm is established for the balance: 'tractionally equal' is to be regarded as an equivalence relation! A functional norm for an apparatus is, in the first instance, an instruction, to human beings of course, to bring about a technically defined state of affairs. Thus, such a norm establishes an aim of technical production and must be supplemented by a system of instructions such that it becomes clear in which manner the desired state can be established.

The first step is to ascertain that the unloaded balance – the balance without the two bodies to be compared being hung on it – behaves symmetrically when traction is applied to the middle rope. When such a balance is employed with a pair of bodies, the symmetrical and transitive character of the traction-equivalence must be maintained by technical means as, for example, in the special case of a balance hanging vertically in a state of rest relative to the earth. Thus, without anticipating mechanical theories, the influence of buoyancy can be recognized as a disturbance of equivalence and can be kept out of the comparison of the two bodies with respect to traction-equivalence (by evacuation of the chamber in which the experiment is taking place, for example). In the special case of linear horizontal movement on the earth, the above functional norm permits the maintenance of equal friction for a pair of dynamically guided bodies by the employment of a pair of equally constructed trolleys. It can easily be demonstrated that 'traction-equivalence' can in fact be realised without knowledge of physics, i.e. without previous knowledge based, in particular, on the measurement of force or mass. There can of course be no proof of this possibility in advance. This possibility is a contingent fact known to us through the history of craftmanship and technology, which is, of course, a historical and not a physical experience.

In this context, one frequently encounters the erroneous view, suggested by an empiricist understanding of theory, that we here have a case of methodological circularity, since empirical scientific knowledge of the possibility of symmetrical balances under real circumstances must already be available to us, in the form of the laws of levers for example. But it is overlooked that in fulfilling the functional norm for a balance we are concerned with actions. Whether actions are possible can definitely be demonstrated by their execution and does not need proof by scientific arguments.

So far we have, admittedly, no measurement procedure for mass. It is rather the case that we merely have at our disposal a two place predicate 'tractionally equal', with a known logical structure (a logical structure of equivalence) which has been artificially established by mechanical means. The transition from traction-equivalence to a metric concept of mass then follows by means of an instruction of the production of homogeneously dense bodies. A body may be regarded as being *homogeneously dense* when any two parts of it which are equal in volume are tractionally equal. This permits us to reduce the metrics of mass to the metrics of volume and length.

Here again we also employ a historically contingent fact, namely, that we know what it means to make a body homogeneously dense. For liquids, it suffices to mix them; technically more important, however, is the production of pure metals and homogeneous alloys.

For any two bodies (for which, of course, no homogeneous density is presupposed) the mass-ratio is equal to the volume of two partial bodies of a homogeneously dense body tractionally equal to these former two bodies. A simple and well-known example of the comparison of mass would of course be to weigh two bodies successively by comparing them with tractionally equal amounts of water, and then to compare the volumes of water in each case, or more simply to work with sets of metal weights, which for methodological reasons have been calibrated with their volume. The *equality of mass* is defined as the mass ratio[18] assuming a measurement process for establishing mass ratios in rational numbers with the help of materially homogeneous sets of weights calibrated according to volume. For logical reasons two bodies are equal with respect to mass if they are tractionally equal. This is of course not the definition of mass equality. It must be emphasized that the common expression 'set of weights' does not imply that 'equal in mass' in the sense of the given definition is a logical equivalent to 'equal in weight'.

Now that mass has been explicitly defined and can be measured without recourse to "inertial systems of reference" or "free motion" we may construct, by means of further steps, a system of mechanics which leads up to the principles of interest here. Even at the level of everyday knowledge it is easy to distinguish free movements from partially or fully guided movements by means of the criterion suggested in the following examples. A stone which has been thrown may be termed *freely moved*; a body directed over a surface, such as a billiard ball on a billiard table, may be termed *partially guided*; the movement of a railway carriage *guided*, or, as expressed above, *kinematically guided*. As a further precondition we now require the measurement of time in the form in which this already exists as an explicitly founded theory in protophysics. We can then go on to develop a terrestrial ballistics, that is, a ballistics formulated within the frame of reference of the earth to account for free motion or projectile motion. Mere methodological consistency would first require kinematics, that is, a theory of the paths and velocities of objects and as a next step a dynamics of the forces affecting projectiles and their resulting behaviour. It is not necessary at this point to attempt the problematical formulation of a principle of inertia or a definition of an inertial observer. Thus the proper name 'earth' only occurs in a special ballistics.

If one wishes to adopt Newton's idea of applying terrestrial ballistics to the heavens, whereby the Keplerian laws of the movements of the planets and the Galilean law of free fall are assumed to describe one and the same sort of motion, one then requires a definition of "inertial systems". Otherwise, one would have to replace the proper name 'earth' by other proper names such as 'sun' or 'fundamental astronomical system of coordinates' and we would thus be faced with the difficulty of not being able to relate our measurements to the new explicit system of reference in a de facto manner but only by way of calculation. If however – and this is not without historical precedent – one wishes to develop a mechanics further, in order to define a system of reference or a class of systems of reference in such a way that, within them, the effects of equal projectiles are also equal we can now, with the aid of the measurement of length, time and mass, obtain the definition. A *system of inertia* is a system of reference in which the law of impact is valid in the case of the ideally inelastic collision.

The above-mentioned difficulties in defining both mass and the principle of inertia no longer arise with this definition. What constitutes an ideally inelastic shock can be just as easily defined by geometrical criteria alone, as can the qualification not explicitly defined above that the shock should be a central one. Thus, we are merely left with the task of measuring the masses and the velocities of the bodies concerned. We have just supplied a definition of mass which is independent of the law of inertia, systems of inertia and a theory of friction.

As in traditional definitions of mass we here assume that relative velocities of bodies which are "free from influence by forces" can be measured. We do, however, have a non-circular definition of "freedom from influence by forces", namely, the law of impact. This law here assumes definitional character resulting from the technical aim of wishing to measure the effect of projectiles colliding with other bodies. Thus, when the collision procedure is repeated, judged on the basis of its ability to 'overrun' another body, the velocity of a projectile must be related to a system of reference in which equal projectiles at equal velocities have equal effects on equal bodies. This demonstrates that to a great extent it is on the basis of technical aims and methodological maxims that apparently explicit physical laws become definitional maxims.

The starting point for the deliberations outlined above was the view, based on an affirmative analytical philosophy of science, that 'mass' has to be a theoretical term which is at least partially interpretable and which relies on a complete theory of mechanics because there is no way of defining mass in a non-circular manner. This view may now be rejected. If we now deviate from the path onto which physics strayed as a result of the famous interpretations of inertial mechanics in the seventeenth century and if one defines mass by means of special conditions of symmetry applied to forced motion by means of homogeneous density and volume, then the mass ratio can, in fact, be defined operatively and in a non-circular way.

With the adoption of this approach, however, the interpretation of the language of physical and more generally, of scientific theories which is widely held today is no longer valid. Instead of disguising the historical lack of explicit definitions by formally demanding philosophies of science it has proved considerably more promising to construct systematically a terminology equivalent to the existing one in physics.

Quite apart from this systematic development of the debate on fundamental concepts of physics the suggested definition permits us to approach Newton's *Principia* in a new way. If we ignore the foreword, Newton begins – without further preliminaries – with a definition. According to this definition the quantity of matter is to be measured by means of density and volume taken together. In his explanatory notes it is stated that the quantity of matter is, in what follows, to be understood as body and mass and that the quantity of matter can be ascertained by the weight of the body in question.[19]

To the modern reader, who is accustomed to defining density as the quotient of mass and volume, Newton's definition seems strange. All of us who follow in the Newtonian tradition have become accustomed to regarding mass as the more elementary concept since it is more frequently used. For Newton the opposite was apparently the case. It is, of course, also the case that pre- and extra-scientifically an elementary – if not necessarily metrical – concept of density is familiar to everyone. Everyone knows, for example, what is meant when, in imprecise everyday language, it is maintained that lead is heavier and wood lighter than water.

What Newton does not explicitly discuss is the question of how density – expressed in modern terminology – becomes a metric concept, that is, how a number of measurement can be assigned to the ratio of two masses by an act of measurement.[20] The definition supplied above can remedy this deficit in Newton's mechanics merely, as it were, by means of prefacing his mechanics with a short introduction. This addition then permits a non-circular interpretation of Newton's three laws of motion.

Once again suitable systems of reference within which laws of motion can be formulated must be chosen in a methodologically justified way. If, let us say, led by technical interests and employing our everyday experience with "heavy" objects with a low friction quotient, for example, bowling balls, we pursue a program restricted to the measurement of acceleratory forces; if, in other words, Newton's second law[21] is to be elevated to the status of a definition of force in a programmatic and technically justified way, then it is advisable to design systems of reference in such a way that acceleration of a body does not take place unless external forces are acting on it. This formulation, which is affected by the same problems of definition as occur in the traditional interpretation of Newton, can be converted into a non-circular one by choosing the above defined systems of inertia as systems of reference.

Let us now assume that the notion of equal behaviour of bodies which are equal in mass in the case of non-elastic collision can be terminologically covered by a definiton of force which is still to be supplied. In such a case Newton's third law[22] can be interpreted as a mere terminological rule of symmetry for the direction and magnitude of the behaviour of two bodies in the case of non-elastic collision. This rule of symmetry could for its own part once again be justified by the ultimately technically motivated decision to conduct comparisons between bodies with respect to geometric, kinematic and dynamic criteria in such a way that neither of the two bodies can be entirely distinguished by means of the comparison procedure itself.

According to this definition of systems of reference it makes sense to interpret Newton's second law as a definition of force, whereby magnitude of motion is already operatively defined as the product of mass and velocity. Within such a constructive framework the first law of motion[23] – the principle of inertia – is a logical implication of the second law: assuming the constancy of mass, the first law of motion follows directly from the proportionality of force and changes of velocity in systems of inertia.

Thus it may be maintained that over and above the systematic achievement of the existing definition of mass Newton's classical mechanics has also been made satisfactorily interpretable to the modern reader. Hence, we may regard the controversy over the definition of mass which has lasted for more than 200 years as being systematically concluded, Newton as being rehabilitated, analytical philosophy of science as having been shown guilty of error, and modern theories of the language of science as having been deprived of one of their most important arguments against explicit definitions.

NOTES

[1] Cf. C. G. Hempel, *Grundzüge der Begriffsbildung in der empirischen Wissenschaft*, Düsseldorf 1974, p. 82 "Die hier umrissenen Überlegungen ... machen es ratsam, ... die Vorstellung aufzugeben, daβ die Sätze einer Theorie in zwei durch erkenntnistheoretischen Merkmale unterschiedene Klassen zerfallen: durch sprachliche Verabredung gesicherte Sätze, die dem Hinweis dienen, was die theoretische Terme bedeuten sollen, und Sätze, die empirische Behauptungen mittels interpretierter Terme ausdrücken und der Bestätigung oder dem Bestätigungsentzug durch empirischen Test unterworfen sind."

[2] J. D. Sneed. *The Logical Structure of Mathematical Physics,* Dordrecht 1971.

[3] Cf. Hempel, *loc. cit.*, p. 65.

[4] Cf. Hempel, *loc. cit.*, p. 58.

[5] The circular character of definitions of mass, force and inertial system of reference in physical textbooks sometimes is hidden behind terms like "neutral test-bodies": "Eine Entscheidung über die Kräftefreiheit eines Massenpunktes kann aber auch unabhängig vom Bezugssystem getroffen werden, wenn es hinsichtlich der vorhandenen Kraftfelder neutrale Probekörper gibt, die von den betreffenden Feldern nicht affiziert werden. Tatsächlich gibt es neutrale Probekörper hinsichtlich aller existierender Felder mit Ausnahme des Gravitationsfeldes." P. Mittelstaedt, *Klassische Mechanik*, Mannheim 1970, p. 41. The same book defines inertial systems in the following way: "Ein Bezugs-

system ist genau dann ein Inertialsystem, wenn ein in Bezug auf nicht gravitative Kräfte freier Massenpunkt sich geradlinig gleichförmig bewegt." Apparently the author does not see that a knowledge about neutral test-bodies or about fields depends on the technical and theoretical availability of mass or force measurement.

[6] Cf. P. Janich, "Trägheitsgesetz und Inertialsystem. Zur Kritik G. Freges an der Definition L. Langes", in: Ch. Thiel (ed.), *Frege und die moderne Grundlagenforschung*, Meisenheim 1975, S. 66–76. Wieder in: M. Schirn (ed.), *Studien zu Frege III*, Stuttgart-Bad Cannstatt 1976, 146–156.

[7] H. v. Helmholtz, *Zählen und Messen, erkenntnistheoretisch betrachtet*, Leipzig 1887.

[8] Cf. P. Janich. "Zur Protophysik des Raumes", in: G. Böhme (ed.), *Protophysik*, Frankfurt 1976, 83–130.

[9] Cf. P. Janich, "Konsistenz, Eindeutigkeit und methodische Ordnung: normative versus deskriptive Wissenschaftstheorie zur Physik", in: F. Kambartel and J. Mittelstraß (eds.), *Zum normativen Fundament der Wissenschaft*, Frankfurt 1973, 131–158.

[10] B. Thüring, *Die Gravitation und die philosophischen Grundlagen der Physik*, Berlin 1967.

[11] P. Lorenzen, "Zur Definition der vier fundamentalen Meßgrößen", in: *Philosophia Naturalis* **16** (1976), 1–9.

[12] W. Stegmüller, *Probleme und Resultate der Wissenschaftstheorie und Analytischen Philosophie II, 72 Theorienstrukturen und Theoriendynamik*. Berlin, Heidelberg, New York 1973.

[13] W. Stegmüller, *loc. cit.*, p. 15.

[14] W. Stegmüller, *loc. cit.*, p. 119.

[15] Cf. W. K. Essler, *Wissenschaftstheorie I*. Freiburg, München 1970, p. 101.

[16] G. Galilei, *Discorsi*, 3. Tag, dt. A. v. Oettingen (Hrsg.), Darmstadt 1964, p. 194, 195.

[17] *Protophysics of Time. Constructive Foundation and History of Time Measurement.* (Boston Studies in the Philosophy of Science, Vol. 30, Dordrecht, Boston, Lancaster 1985)

[18] "quantitas Materiae est mensura ejusdem orta ex illius Densitate et Magnitudine conjunctim."

[19] "... Hanc autem quantitatem sub nomine corporis vel Massae in sequentibus passim intellego. Innotescit ea per corporis cujusque pondus."

[20] I do not deal here with the Newtonian concept of density which – roughly speaking – was intended as the number of atoms in a certain volume. For this concept of density does not open a way towards metrization of mass in a strict operational sense.

[21] "Mutationem motus proportionalem esse vi motrici impressae, et fieri secundum lineam rectam qua vis illa imprimitur."

[22] "Actioni contrariam semper et aequalem esse reactionem: sive corporum duorum actiones in se mutuo semper esse aequales et in partes contrarias dirigi."

[23] "Corpus omne perseverare in statu suo quiescendi vel movendi uniformiter in directum, nisi quatenus a viris impressis cogitur statum illum mutare."

PAUL LORENZEN

ON THE DEFINITION OF 'PROBABILITY'

When applying the theory of probability in statistics one usually spends little time considering definitions of the term "probability". Only in the philosophy of science do we find an ongoing controversy about the concept of probability.

If one disregards the philosophical specialists' terminology (for example, "subjective" and "objective" probabilities) one is essentially confronted with two basic positions: "empiricism" and "apriorism". The empiricist wants to reduce probability to *observed* (if also idealized) frequencies; the apriorist wants to get along *without* observation, and hence without empiricism.

In what follows I suggest a definition which begins with an idealization of frequencies but which adds a normative concept of randomness by taking into consideration the manufacturing standards for random-event generators (*Zufallsgeneratoren*). The law of large numbers is then a *proof* that probabilities are idealized frequencies.

Consider a set (population) of N elements $C_1, \ldots, C_N$, n elements of which satisfy the proposition $A(x)$. The frequency of A – usually called the "relative frequency" – is defined as

$$\rho(A) = \frac{n}{N}$$

On the basis of elementary arithmetic the following theorems hold for frequencies:

(I) $\rho(A) = 1$, if all elements satisfy A

(II) $\rho(A \vee B) = \rho(A) + \rho(B)$ for the disjunctive propositions A, B.

If one defines conditional frequency $\rho(B/A)$ as the frequency of B in the subset of elements with A, then the following theorem holds:

$$\rho(B/A) = \frac{\rho(A \wedge B)}{\rho(A)}$$

If one takes an element c from the population as a "sample" (*Stich-*

Robert E. Butts and James Robert Brown (eds.), Constructivism and Science, 163—170.

probe) one might say that $A(c)$ holds "with the probability $\rho(A)$". Here we are concerned with the prediction $A(c)$. If $\rho(A) = 1$, then $A(c)$ is *necessary*; if $\rho(A) = 0$, then $A(c)$ is *impossible*; if $0 < \rho(A) < 1$, $A(c)$ is *contingent*, that is, possible but not necessary.

Probability theory must give an account of the conditions under which it makes sense to make stronger claims about contingency. When may we add a clause such as: "the more nearly $\rho(A)$ approaches 1, the more a contingency becomes a necessity"? Or, to use a comparative, "the more necessary $A(c)$ becomes"?

To specify a metric, one may introduce $\rho(A)$ as a "degree of necessity" of $A(c)$. Instead of the term "degree of necessity" one usually uses "probability" ("*Wahrscheinlichkeit*", literally, truth-resemblance) which was first introduced as a translation of "*verisimilitudo*". The more "probable" a proposition is, the more similar it is to the (necessary) truth.

However, it makes no sense to increase the precision of the claim that $A(c)$ is contingent for some "sample" to the claim that $A(c)$ has the probability $\rho(A)$ if one holds that *every* taking of an element is a trial. If one *intentionally* takes a c which satisfies $A(c)$, then $A(c)$ is necessary, even if $\rho(A) < 1$. In a *sample*, an element must be *randomly* selected.

To give an account of statements of probability randomness must therefore be defined. This can be done using random-event generators such as dice or wheels of chance. A device may be called a random-event generator if it meets the following conditions:

(1) *Uniqueness*: Every use of the device (every "trial") results in exactly one of finitely many propositions $E_1, \ldots E_m$ (elementary events).

(2) *Indistinguishability*: Causal knowledge provides no basis for distinguishing in advance one of the outcomes $E_1, \ldots, E_m$ from another.

(3) *Repeatability*: After every trial, the device is in the same condition as beforehand.

In developing our account of probability, we make use of the historical fact that technically *good* random-event generators are produced in our culture. There is no "perfect" random-event generator but there are sufficiently good realizations of the ideal norms of uniqueness, indistinguishability and repeatability. If an element c is taken randomly from a set $\{c_1, \ldots, c_N\}$ of N elements (which is now supposed to mean: with the help of a random-event generator) and if $\rho(A)$ is the frequency of the proposition A, then one can account for giving the prediction $A(c)$ the probability $\rho(A)$. A mere definition does not suffice because the probability of $A(c)$ could be defined as $\rho(A)$ even in cases of non-random choice. As we have seen, that would be unreasonable.

One obtains an account of random choice by first assigning for every c_ν a probability ω to the proposition $c = c_\nu$. Because of the uniqueness of the random-event generator "$c = c_1 \vee \ldots \vee c = c_N$" is necessarily true. Thus according to I:

$$\omega(c = c_1 \vee \ldots \vee c = c_N) = 1$$

Due to the uniqueness of the random-event generators, two or more of the propositions "$c = c_1$" ... "$c = c_N$" cannot be simultaneously true. Thus according to II:

$$\omega(c = c_1) + \omega(c = c_2) + \cdots + \omega(c = c_N) = 1$$

Finally, because of the indistinguishability of outcomes in random-event generators

$$\omega(c = c_1) = \omega(c = c_2) = \ldots = \omega(c = c_n)$$

Thus, for all $\nu = 1, \ldots, N$

$$\omega(c = c_\nu) = \frac{1}{N}.$$

If A has the frequency $\rho(A)$ in $\{c_1, \ldots, c_N\}$ then there are $\rho(A)$ times N elements c_ν with $A(c_\nu)$. The probability that c is one of these c_ν is according to II, therefore:

$$\rho(A) \cdot N \cdot \frac{1}{N}$$

That is, $\omega(A(c)) = \rho(A)$.

In order to arrive at this equation, we have postulated for all propositions A about the outcomes of a random-event generator used in the calculation of a probability ω:

I. $\omega(A) = 1$, if $A(c)$ is necessary

II. $\omega(A \vee B) = \omega(A) + \omega(B)$, if $A(c)$ and $B(c)$ denote disjoint outcomes.

III. $\omega(E_1) = \omega(E_2) = \ldots = \omega(E_m)$

For simplicity's sake we have followed the usual practice of writing "$\omega(A)$" instead of "$\omega(A(c))$".

I and II are founded on the corresponding theorems about frequencies; III is founded on the indistinguishability of outcomes for random-event generators.

Before carrying out a trial with a random-event generator, every proposition about an outcome $E_\mu(c)$ for $\mu = 1, \ldots, m$ is contingent. But due to the indistinguishability of outcomes we have here a special type of contingency. What is special is that here, as in roulette, a wealth of causal knowledge is used to *produce* machines such that (according to our best causal knowledge) no distinction between possible outcomes may be made before they occur. That "red" is "more necessary" than "black" may *not* be maintained, simply because of the construction of the wheel. The indistinguishability requirement is a requisite for the manufacturer of random-event generators. More is known about $E_\mu(c)$ than that it is merely contingent. It is suggested that the propositions $E_\mu(c)$ be called "*totally contingent*" or in short "random" (zufällig). Using this terminology, random-event generators are devices whose outcomes are *random*.

Suppose that a series of L trials is carried out using a random-event generator. For every disjunct of the series $E_1, \ldots, E_m$, two calculations should be made. On the one hand, the probability $\omega(A)$ may be calculated using I–III; on the other hand, the frequency $\rho_L(A)$ of A in the series of L trials can be determined. Further, for any positive ε, a probability ω_L of $|\rho(A) - \omega(A)| < \varepsilon$ can be calculated. This proposition $|\rho_L(A) - \omega(A)| < \varepsilon$ is thus a proposition about any *arbitrary* series of L trials. There are m^L such trial series. $\omega_L(|\rho_L(A) - \omega(A)| < \varepsilon)$ is the frequency of the series of L trials which satisfy the proposition $|\rho_L(A) - \omega(A)| < \varepsilon$. (That this frequency is the desired probability follows from the fact that all m^L series of trials – or more exactly, the m^L outcomes of a series of L trials – have the same probability. Because of the repeatability condition, these m^L outcomes are indistinguishable.) According to Bernoulli, the probability $\omega_L(|\rho_L(A) - \omega(A)| < \varepsilon)$ converges to 1 for $L \to \infty$. This is the "weak law of large numbers":

$$\lim_{L\to\infty} \omega_L(|\rho_L(A) - \omega(A)| < \varepsilon) = 1$$

I's significance is as follows: the statement that the frequency $\rho_L(A)$ is the probability $\omega(A)$ within a certain margin of error ε becomes a practical necessity as L gets very large. Before Bernoulli's theorem one was not warranted, when considering propositions about trials with random-event generators, in calling a number calculated according to the frequency-propositions I and II a probability.

For continuous random-event generators such as gambling wheels there are no elementary events. Indeed, the wheel comes to rest in some position but it would be absurd to assign a positive probability to every "point". Instead, the circumference of the wheel is divided into arbitrarily many equal segments and each of them gets the *same* probability. If the circumference had the length 1, then the probability of any interval would simply be its length. For a random-event generator which comes to rest somewhere on a rectangle (with area 1), the probability that the device comes to rest within a given subdomain [*Teilfläche*] would, under corresponding conditions, be defined as the area of the subdomain. This definition is exactly as "arbitrary" or "reasonable" as defining "area" itself as the limiting value of the areas of the subrectangles.

Stating II more precisely, we get "full additivity" for every series A_* in which no pairs $A_{\nu 1}$, $A_{\nu 2}$ occur together (*mit paarweise inkompossiblen* $A_{\nu 1}$, $A_{\nu 2}$):

$$\text{II}\sigma \qquad \omega(V_V A_V) = \Sigma_V \omega(A_V)$$

Using the results of modern measurement theory (Borel, Lebesgue, Fréchet) Kolmogorov (1933) was able to show that the definition of probability always leads to a "probability field" – in set theoretic terms, to the defining of a normed fully additive measure on a σ set-field. The discrete random-event generators lead to Laplacean probability fields, the continuous to Lebesguean probability fields in n-dimensional number space. Following von Mises, it can also be shown that combining random-event generators into random-event aggregates (*Zufallsaggregate*) leads likewise to probability fields in every case. Of course, the fields are in general no longer Laplacean or Lebesguean.

These combinations of random-event generators correspond to operations that may be carried out on probability fields. For *one* probability field, these are above all the relativizings ("partitions" for von Mises) where the transition is made from $\omega(A)$ to $\omega(A/B) = \omega(A \wedge B)/\omega(B)$ for a set B with $\omega(B) > 0$, and the coarsenings ("mixings") with which the set fields are homomorphically mapped onto other fields. Every urn containing balls of different colors affords such a coarsening in that the balls of the same color (that is, their indices) are comprised of sets. A "loaded die" delivers a probability field by coarsening a Lebesguean field. We reduce the problem to two dimensions when we "throw" a square column instead of a cube:

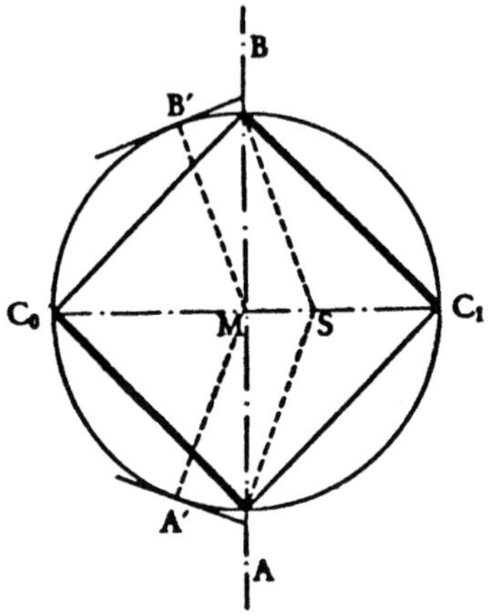

Let the center of gravity S be different from the center M. Let us imagine that the column C_0AC_1B is contained in a cylinder with circumference 1. The column is "thrown" by a random-event generator in such a way that it falls vertically and without angular momentum onto a table and that – were it cylindrical – it would land with equal probability on any of the equally long segments of the (imagined) circumference. Without the cylinder, the column will land on one of the four sides. We want to know the probability that C_0 will land face upwards. If A' and B' are defined such that $A'M$ is parallel to AS, and $B'M$ to BS, then the length of the circumference $A'C_1B'$ is the probability we seek. If the position of S is known, the probability that C_0 will face upwards is calculated to be $\omega > 1/2$.

If one hasn't determined the center of gravity beforehand, one must proceed according to the Bernoulli theorem, using the frequency ρ_L of C_0 in a sufficiently long series of L trials. One finds only the "well-supported" hypothesis $\omega = \rho_L$ about the probability ω. That is, the frequencies $\rho_L \pm \varepsilon$ would have a probability very close to 1, even if ε were small.

More important than relativizing and coarsening is the *product formation* ("combining") of several independent random-event generators. Only with this relation do we get from generators to aggregates. Here independence – like repeatability for generators – can be defined as technically realizable causal independence.

According to our best causal knowledge, no trial with one aggregate effects a change in another. Mathematically, the product-formation of probability fields leads to a new probability field, the product field. From a purely mathematical standpoint, denumerably infinitely many fields may be multiplied. This affords – as in the case of the strong law of large numbers – the comfortable possibility of talking about finitely but arbitrarily many fields.

Kolmogorov showed further that the course of a stochastic process (which at any time can only be determined by a probability field) can also be treated as an event in a probability field. Since Kolmogorov, therefore, the mathematical theory of probability has been practiced as a theory of σ-set-fields with a "probability" of which only theorems I and IIσ have been axiomatically demanded.

I would like to suggest being more cautious and naming the arbitrary models of Kolmogorov axioms 'Kolmogorov fields'. All models that we obtain when beginning with Laplacean and Lebesguean fields, we apply all operations (relativizing, coarsening, product-forming and process-forming), fulfill additional axioms. In all applications we are concerned with Borel-fields in topological spaces. It has been shown that the class of so-called "Polish" spaces (which are separable and completely metrizable) is sufficiently large to be closed with respect to all of the mentioned operations on probability fields.

The class of "Polish" Kolmogorov fields is however – in contrast to the class of all Kolmogorov fields – provably not too large: all of these fields can be represented as limiting values of discrete probability fields. The concept of limiting value in use here was specified by H. Cartan: he named it "vague convergence". This approximation theorem justifies simply calling models of normed σ-additive measures of Borel fields in Polish spaces "probability fields".

The relevance of the approximation theorem arises naturally from the applications.

For the applications, one is always pointed toward a *definition* of the concept of probability. This situation is disguised by the fact that in statistical applications (which are much more important that the original applications to games of chance) the random-event aggregates do not appear explicitly but are only hypothetically assumed. As the inertial masses of the sun and the planets are not "theoretical quantities", neither are the probabilities in statistics "theoretical quantities". Rather, people work with the fiction that the observations are the outcomes of unknown (hidden) random-event aggregates, "as though God played with dice". For example, the decay of radioactive material, which can be described as an exponential decay according to the observations, is "explained" as follows. The frequency of decaying atoms is proportional to *dt*. Here *X* is the number of atoms and *dx* (infinitesimally formulated) is the number of decayed atoms during time interval *dt*:

$$\frac{dx}{x} \sim dt$$

From this it follows by integration that $\Delta \lg x = -K\Delta t$ for a positive constant K, that is, the exponential law:

$$x = x_0 \, e^{-K(t-t_0)}$$

The frequency of decay $K\Delta t$ in a sufficiently short time interval Δt is similarly "explained" on the other hand by the *probability* that any atom decays in Δt is $K\Delta t$. Then it is maintained that the decay of atoms happens *as though* a random-event aggregate accounted for the decay.

How statistics (in physics or in other sciences) solves the problem of such statistical hypotheses, and especially how it solves the problem of their isolated verification, is not a part of the philosophical problem. That problem is first and foremost to find a definition for the fundamental concept "probability" and then to account for the application of the definition.

SECTION III

CONSTRUCTIVISM AND THE VALUE SCIENCES

OSWALD SCHWEMMER

PRACTICAL REASON AND THE JUSTIFICATION OF NORMS. FUNDAMENTAL PROBLEMS IN THE CONSTRUCTION OF A THEORY OF PRACTICAL JUSTIFICATION

I. THE PROBLEM OF PRACTICAL JUSTIFICATION

The question whether reason is or can be practical, and accordingly whether our practice is or can be made reasonable is, on the one hand, as old as all intellectual efforts towards an orientation of our life. On the other hand, in the form in which I shall discuss the matter, it expresses a particularly modern problem.[1] We find at the beginnings of our philosophical tradition – in the fragments of the Pre-Socratics, in the works of the Sophists and in the "classical" Greek philosophy of Socrates, Plato and Aristotle – the theme of the good life and just conduct at the centerpoint of their inquiries. But the question of practical reason as an inquiry into the *First principle* of a reason-grounded practice arises as such only when the demand for the legitimation of the imposed norms of common life attains a position of importance in itself. The critical occurrence is then no longer, as in antiquity, only the relief of a divinely grounded legitimation by a legitimation under an obligation to the common responsibility of the citizens. It is rather the act of legitimation without a common basis, without a mutually imputed understanding of life and actions.

It should be stated here that the philosopher's question, contrary to general opinion, and in contrast with the Hegelian dawn-flight metaphor of Minerva's owl, did not represent a richly reflected post-script to an aging epoch, but rather an albeit abstractly conceived anticipation of an epoch in the making. For the loss of a common basis for the legitimation of our lives and actions had not taken place by 1789. The fanfares of equality, freedom and brotherhood were proclaimed by an awakening citizenry which established a position of political equality with the aristocracy or the old regime, but which did not (even as a propertyless class) seek to legitimize the possession of property as a right in itself: economic legitimation was given neither a

Robert E. Butts and James Robert Brown (eds.), Constructivism and Science, 173–190.

superordinate nor even equal footing with its political counterpart. The disappearance of the situation in which this equality, namely a common responsibility for that which was at least partially inherited, did not come about until the middle of the following century (around 1850) when both Hobbes's *Leviathan* (1651/1688) and Kant's *Critique of Practical Reason* (1788) had long since been written and partially forgotten.

The reference to the historical development of the demand for legitimation of those norms governing our actions shows two things. First the demand for legitimation requires an enquiry into the First principle or principles of practical reason, that is, of the justification of our actions, only when a material norm or a fixed value as a basis for legitimation or justification is no longer at our disposal. *Ethics* as a question of practical reason, or of practice grounded in reason, becomes thereby the *question for the principles of practical justification*; every material norm is in principle made subject to the demand for justification, and this occurs in principle when *unlimited demand for justification becomes a social reality.* This reference shows furthermore that ethics is then subjected to a rationalization and relativization of its problems. For without a basis for justification (that is, without material norms or values) which stands on its own grounds, specified tasks of justification (in the sense that specified actions or norms are to be justified) are no longer fixed a priori. The task of justification is determined by its social reality. Only when the demand for justification has found a (real existing) subject, that is, when the actors demand in a socially effective manner a justification for the norms imposed on them, only then does the question of the justification of these norms arise, because only then does an addressee for the justification exist. *Ethics* is, accordingly, no longer defined by a fixed catalogue of specified tasks of justification or by specific norms or values but becomes rather a *methodology*, whose sphere of application is not determined from within itself but rather from the difficulties existing between the authors and the addressees of norms. *Ethical themes are relative to those* factual and (to clarify their social relevance) *historical instances in which they are handled.*

I am aware that I am taking up a modern question by formulating the question of practical reason as a question of the principles of practical justification – and this not for the justification of previously specified and relevantly defined norms or values but rather for those actions and norms for which a need for justification has evolved from certain situations. And I am handling the theme of the good life (although on the basis of its historical development) within the *theoretical aspect of justification*. Thus practical reason will not be treated in the breadth with which it was handled by Socrates, Plato and Aristotle. But the problem of practical reason assumes a critical character as the problem of practical justification, a formulation which Greek philosophy did not deem necessary but which we, since Hobbes and Kant, are forced to deal with.

In the following text I shall discuss the practical problem of justification as the question of the principles (or the First principle) of practical justification in the threefold manner: (1) as the question of the *methodical construction* of practical principles of justification; (2) as the question of the *pragmatic justification* of these principles; and (3) as the question of the *material application* of these principles.

II. THE METHODICAL CONSTRUCTION OF THE PRACTICAL PRINCIPLES OF JUSTIFICATION

I shall begin this methodical construction of the principles of practical justification with a mere formal construction of *types of propositions* and of relations between these types of propositions, namely, specific *sequences of types of propositions*, which I will term *justifications*. By explicating the rules and principles employed in constructing such sequences I shall reveal, so to speak, the *logic* and *syntax* of practical justification. Thereafter I shall supply the methodical justification of this formal construction.

1. *Formal Construction*

When we speak of practical justification, we are referring to the justification of our actions as opposed to the justification of assertions. This distinction avoids a misunderstanding which has led to a not inconsiderable neglect of methodically orientated thinking in the sphere of practical philosophy. It is clearly the case that every justification is a *justification of propositions* and that every utterance of such propositions is an *action* – as well as being the assertion of statements – and is therefore subject and accessible to justification. The propositions however, at which practical justifications are aimed, are not statements but exhortations. At this point I shall adopt the grammatical terminology and refer to all exhortations, whether they appear in the form of advice, suggestions, commands, favors, etc., as *imperatives*. The decisive difference *vis-à-vis* the statements – the affirmative propositions or indicatives – is that imperatives do not assert the existence of a state of affairs, but that by complying with them the occurrence of a state of affairs is intended to be brought about (whereby the meaning of "intended to be" can be relativized according to whether the imperative is advice, a suggestion, etc.). At the end of those sequences of types of propositions, which may be formally classified as justifications, there must inevitably be imperatives. The question now is – and this is the essential question of the formal construction of practical justifications – which sequences of types of propositions ought to lead to such imperatives.

(a) *Finalization*: The first step in justification is made by distinguishing between afinal and final imperatives. I shall term those imperatives *afinal* which call for the performance of an action, and those imperatives *final* which call for the bringing about of certain states of affairs by an action. This distinction (which, as I shall propose it, is prior to the distinction between means and ends) is basic to a language theory as well as to a justification theory. By enabling us to represent our actions as a cause of specific states of affairs, this distinction also enables us (in the scope of language theory) to distinguish between our actions according to the states of affairs which ought to be brought about by them: a series of uninterpreted body movements is defined as a meaningful action by the final imperative joined with it. At the same time, it is this distinction which enables us (in the scope of justification theory) to justify actions at all. Because actions are distinguished as meaningful units on the basis of the states of affairs they are intended to bring about, they also are justified as meaningful actions by just these states of affairs. To learn the distinction between afinal and final imperatives is at the same time to practice the first step of justification, namely, the step from final to afinal imperatives. In this first step an action or an afinal imperative is justified precisely when and to the degree that by this action a specified state of affairs is brought about. The action or the afinal imperative calling for its performance is then *justified relative* to this state of affairs, or to the final imperative calling for its being brought about.

It should merely be noted that the distinction between afinal and final imperatives is only meaningfully applicable if the state of affairs, the bringing about of which is called for in the final imperative, could not also occur without a particular action, that is, if the performance of an action is a condition of (= necessary for) the occurrence of this state of affairs. Thus it follows that an action A, or the imperative $!A$ calling for its performance is only justified if it is true that S – which is the state of affairs, the bringing about of which the justifying final imperative calls for – occurs only if A is performed. (A here can stand for a class of different actions, one of which must be performed so that S occurs.) The following formula thus expresses a formal condition that the final imperative $!S$ justifies the afinal imperative $!A$: $R \wedge \neg A \rightarrow \neg S$, whereby R stands for the actual existing ("real") situation.

In everyday language we frequently express this relationship between afinal and final imperatives by the use of "in order to": one performs this action in order to bring about that state of affairs. Lest there should be a misunderstanding of the above-proposed reconstruction of a first step of justification as a description of our everyday use of "in order to", I shall complement my logical-syntactical remarks with a remark on the semantics of imperatives.

The distinction between afinal and final imperatives is practiced with the actions the afinal imperatives call for and which are at the same time justified by the final imperatives (as the bringing about of a certain state of affairs). This acting is always *interaction* – otherwise we would not need imperatives: several persons exhort one another to perform actions and to bring about states of affairs and accordingly to comply or refuse to comply with these exhortations. In such interactions the actions and states of affairs represented in the imperatives must be – in principle – equally comprehensible to each of the participants, that is, they must be represented in a language that can be verifiably commonly practiced. Without outlining at this point a theory of a methodical language construction I suppose that this condition of common practicability or – equivalently – teach- and learnability is fulfilled at least by those distinctions which can be introduced in an *exemplificatory* manner; which can be practiced with examples and counter examples; which can be referred to in common contexts of interaction. As long as one is concerned only with non-linguistic actions – such as going, bringing something, opening a window, etc. – both the descriptions of the actions and the representations of the states of affairs in the imperatives can only be given by means of exemplificatorily introducible predicators (words we use to represent our distinctions).

When I talk of the justification of an afinal imperative by a final imperative, I am not suggesting that one could perhaps justify certain actions by stating that something – as a result of them – is rendered better or more just. It is precisely this which remains to be ascertained and for which we still lack sufficient methods of justification. The exhortation to establish a just society is not a final imperative in this sense. No specific state is determined by this exhortation; that would only be the case if one were to add a specification of the forms of organization for a society. And they would still need to be judged as to their justness.

I mention this, so that in my formal considerations the connexion between the formal rules and their content, that is, the real things we speak about, is at least hinted at.

(b) *Generalization*: How can final imperatives, for their part, be justified? A first answer, which at the same time introduces the second schematic practical step in justification, consists in referring to the final imperative as the *case of a rule*. If we remember that the states of affairs, the bringing about of which the final imperatives call for, are representable exclusively by exemplificatorily introducible predicators, then we see that the corresponding final imperatives are not and cannot be complied with always and everywhere. Indeed, one can strive always and everywhere for justice and truth, but one cannot at every moment of one's life exhort oneself or others to bring about or maintain a specified organizational state. Final imperatives are, therefore, conditional imperatives, that is, they are not simply proposed but rather proposed relative to the existence of certain situations (even if the situation is not explicated when the respective imperative is proposed). Some imperatives are proposed regularly, others are not. The regular conditional imperatives are, at least for the person who proposes or complies with them, a reason in the sense that one can refer him to the rule: "For each situation (of the type) *R* you have proposed the imperative to bring about the state of affairs *S* (or you have acted as if you had proposed or complied with this imperative !*S*). Now such a situation (of the type) *R* exists. So propose the imperative !*S* (to others) or comply with it (yourself)!" This exhortation to consistent compliance with the rules supposes, if interpreted as a justification, that a certain constancy of our acting is as such rational (even if, of course, the different specified rules we have proposed or complied with cannot yet be supposed to be rational). This exhortation does not, however, suppose that a rule, once chosen, can always be employed as a reason for an individual imperative. The changing of rules can also be justified – as will be shown in the next step. Only if someone *asserts* (here and now) that he is following a rule or if he (here and now) *proposes* a specified imperative for specified situations (and if and as long as someone acts as if he had made this assertion or proposed this imperative), then one can refer to this assertion or exhortation as a reason for an individual imperative. Were there no regularity at all in human actions and speech, one would not be able to justify any final imperative. One might ask, however, whether such human beings could be regarded as rational if, so to speak, every breath of change altered their actions and speech.

Regularity in complying with or in the proposing of an imperative shall be termed its generality. Thus the second step in justification consists in a *generalization*, namely, in that a conditional general final imperative is sought which is applicable to the existing situation and from which it can be concluded that the (individual final) imperative to be justified is to be complied with.

Here I would also like to add a terminological remark. A regular compliance with or proposal of specified final imperatives also makes possible the regular attribution of these final imperatives to certain afinal imperatives of the acting person – or simply to his actions. In different situations one can attribute certain final imperatives to one's actions. In particular even when the final imperative has not been explicated, one can nevertheless assert that one's action shall bring about precisely the state of affairs which is called for in this final imperative. I shall term such an assertion an *interpretive assertion* – or an interpretation, for short. An action is interpreted by the assertion that the action complies with the imperative which calls for bringing about the state of affairs *S*. The bringing about of this state of affairs I shall term in this case the *end* of the action. The performance of this action is termed accordingly a *means* to this end.

Since their use requires an interpretation, the terms 'end' and 'means' are no longer introducible exemplificatorily and thus do not belong to the language by means of which, in the first stage of the practical justification, actions and states of affairs are represented in the imperatives. If one terms the language consisting of the exemplificatory introducible predicators *observation language*, without thereby associating the Carnapian distinction between observation and theory language, then 'end' and 'means' are the basic terms of the *interpretation language*. So the introduction of a new language level is semantically linked with the logical-syntactical addition of generalization as the second step of justification.

(c) *Universalization*: If we succeed in justifying certain actions with general imperatives, the suggestions made so far still do not say anything about how one is to come to a justified decision, when the general imperatives lead to mutually incompatible exhortations, that is, to exhortations where compliance with one excludes compliance with the other. But such situations are decisive in the problem of practical justification. I shall term them *conflict situations*. The problem they raise in the context of justification can be expressed in the question, "How can we justify which general imperatives we should comply with and which we should not?" For it is obvious by definition that not all general imperatives can be complied with; one must therefore find a way – if one wants to act at all – to alter them in such a way that compliance leads to mutually compatible exhortations.

In the first and second step the justification of an action (or of the corresponding afinal imperative) could be accomplished by referring back – reconstructing the steps of finalization and generalization – to exactly those imperatives which the actor himself had proposed or complied with. But in a conflict situation the imperatives of the actors are no longer unproblematically available for justification. We now have the problem of how to come to a justified decision for or against them. The answer, however, is to be found in the question with which I formulated the problem of justification. It is precisely because no imperatives (which are de facto complied with) are available for the justification of actions – and the question of justification is therefore posed in a new way – that one can characterize an imperative which, even in a conflict situation, can be employed to justify an action. It must be an imperative acceptable to everyone and therefore *valid for (obligatory for) everyone*. Expressed in a pointed manner: as *no single person's imperative* can be simply taken as a reason, one has to find an imperative that can be regarded as *everyone's imperative*, that is, an imperative that – in the relevant situation – can be proposed and complied with by everyone.

I shall term an imperative which is valid for everyone in this way a *universal imperative*. In a conflict situation the task of justification therefore consists in universalizing the proposed general imperatives. In reference to Kant I shall term the general final imperatives *maxims* and the universal general final imperatives *norms*. Someone acts justifiably in a conflict situation when he is acting according only to those maxims that can at the same time be valid as norms.

This formulation – a conscious reminder of Kant's Categorical Imperative – represents the third step of justification – the universalization of our maxims. If the three steps are compared with one another, one can say that in the first step the finalization of the afinal imperatives or the actions called for in these imperatives provides a justification which is *rational with regard to the ends*; the second step – the generalization of the final imperatives – proposes, as I would like to express it following Max Weber, a justification that is *rational with regard to the (general) sense (of our acting) at all*. (I shall take the rationality with regard to the ends as a part of the (more general) rationality with regard to the sense.) Finally, the third step presents the specifically *moral* justification. Taking up the philosophical tradition one also can say that the rational justification – that is, the justification that is rational with regard to the ends and to the sense of our actions – makes actions rational ("*verständig*"), while the moral justification makes actions reasonable ("*vernünftig*"). Because of this distinction I shall term the rule of moral justification, the *principle of reason*: "Act in such a manner that the maxims guiding your actions can at the same time be valid as norms!"

In using the terms "rational" (or "right with regard to the ends or the sense") and "reasonable" (or "morally right") predicators of a third language level are introduced. These predicators are characterized by their being used to evaluate (according to certain rules) the exhortations formulated in the observation and interpretation language (that is, on the first and second language level). I shall term such predicators – in so far as their use is explicitly determined by specified rules of evaluation – *evaluation terms*. (In the absence of rules, that is, without an explicitly terminological definition, we have merely evaluation predicators.) Thus we must distinguish three language levels in our considerations of a theory of justification: that of the observation language; that of the intepretation language; and that of the valuation language.

By means of the logic of quantifiers the transition from the steps of rational justification to those of moral justification, moreover, also can be clarified. The generalization of final imperatives, so that we obtain maxims, consists in our no longer speaking about individual states of affairs, which are represented by elementary sentences, but rather about (general) states of affairs which are represented by sentences with variables substituted for the nominators. We quantify with the universal quantifier over the range of values of these variables. The addressees of the imperatives will be defined by an index of the imperative sign: 'p' for a name (or a "nominator") of a certain addressee, '*p*' for the corresponding nominator variable. A general conditioned final imperative – that is, a maxim – then will get the notation:

$$\forall x\ (R(x) \rightarrow\ !_{\mathrm{p}}S(x)).$$

The universalization now consists in the quatification over the range of values of the variable for the person *p*. Thus a norm – that is, a conditioned universal general final imperative – will get the notation:

$$\forall p\ \forall x\ (R(x) \rightarrow\ !_{p}S(x)).$$

Assuming sufficient clarification of this logical distinction between generalization and universalization we can also give a notation of norms and maxims by means only of the logic of connectives. In its simplest form we then can write a norm as $R \rightarrow\ !S$; a maxim gets the additional index for the addressee behind the imperative sign $R \rightarrow\ !_{p}S$.

2. *Methodical Justification*

The first two steps of justification – the rationality with regard to the ends or to the sense of our actions – result from distinctions that we need to cope with life in general, that is, independently of our special goals. As I have already stated, it is by making the distinction between afinal and final imperatives or, as we might also put it, between actions and their intended results that we are able to talk about justifying our actions at all. And the practicing of this distinction already *is* the rational, that is, rational with regard to the ends, the justification of actions. Thus rational justification and distinguishing between actions and ends or between afinal and final imperatives are merely two sides of the same rational doing.

The generalization of the final imperatives results with the end-means-distinction in the same way. The practicing of this distinction makes sense in that it makes it possible for us to plan our actions. We call our ends to mind in order to then find appropriate actions to reach these ends. The generalization of our ends, however, is connected – as a condition of its sense – with the planning of our actions. For only if we want to comply with the final imperatives, which we have now also proposed for future situations, does such planning make sense.

Unlike the steps of rational justification the universalization of rational imperatives does not yet arise from and (simultaneously) with the fundamental distinctions that we need in order to cope with our life. Nevertheless one can justify this universalization in one short step. For if one wants to overcome conflict situations by *justification* at all, then the principle of reason, that is, the demand for universalization or also (as I read it) Kant's Categorical Imperative, indeed appears as the highest principle of practical justification. The decision to overcome conflicts by means of justification, that is, the *decision in favor of and against violence, in favor of the attempt of independent change of will and against force*, this decision hence becomes the basis of the methodical justification of our formal constructions. I shall therefore term this decision the *practical basic decision*.

In speaking of the practical basic decision we touch upon the central theme of philosophical efforts in the realm of ethics. I shall take up these efforts in a twofold manner, with regard to the *terminological* as well as to the *theoretical* problem. To begin with, the question arises as to how we can even formulate this basic decision. The *two* formulations chosen above – revealing our mistrust of the clarity of only one formulation – are by no means terminologically regulated. So how might we terminologically regulate the formulation of this decision? Examples and counterexamples are not available as they are in exemplificatory introductions. One might be tempted, all the same, to work with examples; perhaps by portraying how a conflict situation is overcome by the participants speaking (proposing, objecting, questioning, answering, voting, etc.) on the one hand and by employing physical force on the other. However, if we want to avoid circularity in the introduction of the required terminology, evaluation terms are not yet available to represent these examples or counterexamples. The decisive distinctions in the examples cannot be formulated (in contrast to the distinctions in the counterexamples: what force is we can state at the beginning of a theory of justification, but we can not yet distinguish every – relevant – kind of force!). The evaluation terms are first introducible by means of the principles of justification or, at the earliest, after the formulation of the basic decision. Hence, if we portray the settling of a conflict without such evaluation terms, we perhaps can state that this settling has been successful after a discussion among those concerned. But the utterances in this discussion could have been of a threatening, persuasive, unintelligible or erroneous, etc., nature. For a discussion situation to serve as an example for the participants who have complied with the practical basic decision, we require evaluation terms, which, for methodical reasons, we do not yet have at our disposal.

If we now accept that the practical basic decision is an exertion which has to be performed by the acting human being and which cannot be described as a simple fact but just consists in *not* leaving the facts to themselves, it is, nevertheless, possible to make this decision in a certain sense intelligible. One can, namely, portray the counter-examples, the reduction to the bare facts, the *absence* of this decision, and that is the employment of force in the resolution of conflicts. Looking at the different specific forms of force that we must portray we then can try to delineate certain ways of averting force. The second part of this introduction is no longer mere description but contains proposals.

I shall term this method of introducing distinctions with which one can formulate the practical basic decision a *historical introduction*. It is the introduction by way of reference to the history of force and at the same time by way of attempts to delineate its historically realized or merely imagined averting or overcoming. This introduction must remain as fragmentary as the averting and overcoming of force – as "reason in history" – itself. It is the construction of the principles of practical justification which is an attempt to make it explicit.

Is it possible then – and this is now the theoretical problem – to justify this basic decision for its part? That a methodical justification is not possible is self-evident. For we do not have at our disposal a reason which is methodically prior to this decision. The practical basic decision is defined as such precisely by this methodical priority. But can we find arguments for it which could bring the actors to this decision too? Can we – as I shall term such manner of argumentation as is defined by reference to its results, that is here, to its adoption by the actors – *pragmatically justify* this basic decision?

III. THE PRAGMATIC JUSTIFICATION OF PRINCIPLES OF PRACTICAL JUSTIFICATION

The task of pragmatic justification of the principles of practical justification – that is, of the basic decision – can be summarized in the question: Why should we be moral? Following Kant this question has become one of the central themes of philosophic considerations. In particular two traditions of answering this question have become established: the *transcendental-philosophic* and the *skeptical decisionistic* tradition. The transcendental-philosophic position, which has in present-day Germany been adopted, for example, by Karl-Otto Apel and Jürgen Habermas, is based on the opinion that someone who contests or even doubts the justification of the basic decision must already have made the basic decision. Otherwise he could neither contest nor doubt it. But this is, if it is read as a logical methodical argument, wrong. For if someone thinks the basic decision to be unjustifiable, he does not need to have taken it when he occasionally brings it forward as an argument *vis-à-vis* his discussion partner or if his arguments in this discussion presuppose its good sense (its being rational or reasonable). For he can use it even then, as one means amongst many, in order to achieve his ends. If he, in fact, does then employ it, it is not as a basis for the principle of reason but as a means to his actual ends.

The skeptical decisionistic position, adopted in the German-speaking world, for example, by Hans Albert and Hermann Lübbe, thinks the pragmatic justification of the basic decision and hence the principles of justification in general to be impossible. It can only be left to the individual without any further attempts at justification. The transcendental-philosophic method of argumentation is thus rejected, correctly as it seems to me. But, the skeptical decisionistic position puts itself in a hopeless situation – together with its counter position – by virtue of the question it asks. The decisive question is not how a pragmatic justification of the practical basic decision or the principle of reason is *possible* but rather in what sense it is at all *necessary*.

The problem of practical justification only arises with such acuteness (which leads methodically to the formulation of the principle of reason) where the reference to common life and actions no longer offers a basis for justification and hence on principle each imperative is obliged to be justified. When the demand for such justification becomes universal, moral philosophers are, in a certain sense, relieved of the burden of justifying the principle of reason: for because and in as far as the demand for justification has become universal, it is rational to justify the proposed imperatives universally. As human beings (as the receivers of exhortations) have made themselves subjects who insist on the justification of demands made on them, the moral justification of the proposed imperatives is rationally justified for those who exhort others.

Reason now seems to have been made thereby mere rationality with regards to the ends. This is, however, only true with one qualification. Namely, the demand for the justification of the basic decision (or of the principle of reason) more precisely represents the demand for justification according to certain *rules* of justification – which must first be set up – and hence is a demand for a *rational* (with regard to certain ends regulated) justification, and compliance with this demand can also only rely on rationality (with regard to the ends). *The fact that* the pragmatic justification can be accomplished in this (merely) rational way shows that every effort towards rational justification is subsequent and supplementary to the self-realization of reason attempted in the course of history, to the *fact of reason* as Kant formulates it, which we can reflect upon but can never construct. Bringing someone to the practical basic decision or to compliance with the principle of reason is no concern of a – only subsequently clarifying – terminological construction or of theoretical argumentation but rather of *enlightenment* in the Kantian sense, which clarifies someone's own interests as well as teaching us to give consideration to the interests of others.

IV. THE MATERIAL APPLICATION OF THE PRINCIPLES OF PRACTICAL JUSTIFICATION

Of fundamental importance for every theory and hence also for the theory of practical justification is the question of the applicability of the general proposals which have been made. How are we to arrive at norms – that is, at maxims which can be complied with universally – in conflict situations defined precisely by the incompatibility of the proposed maxims?

The practical basic decision can also be understood as the decision for the subject status of others, that is, the decision – before their critical examination – to first treat the ends of others in the same way that one treats one's own. It is this equal treatment of one's own and others' ends that is demanded by the principle of reason. For this principle permits as reasons only universal imperatives, that is, imperatives which are to be valid for everyone in the same way. The equal treatment demands that one only proposes such arguments which have the ends of the others also as their material basis. In other words: I can only bring someone to change his ends if I show him that this in some way serves his ends. This however seems to be a contradictory task.

This task is not contradictory only if one can establish a rational relationship between the various ends or maxims of an actor which we can use for justification. Such a relationship must be one that allows us to waive certain ends which are the means to other ends in favor of these superordinate ends. I shall not attempt to more closely define this subordinate and superordinate relation between ends and maxims and to further justify the use of this relation. But I shall establish the postulate (1) to seek in conflict situations the mutually compatible maxims, which are superordinate to the mutually incompatible and de facto proposed or realized maxims. (2) To propose new maxims which are equal to the former ones and subordinate to the compatible maxims but which are also – and these are no longer equal to the former ones – mutually compatible. The possibility of proposing such compatible new maxims defines the *general condition of applicability of the principle of* reason. The application of the principle of reason thus supposes the ascertainment and argumentative use of the rational relationship between the proposed or realized maxims of the actors. In other words: The realization of reason requires rationality.

This general condition of applicability for the principle of reason now also supplies a *basis of application*: we are successful in stating some maxims as norms because they are rational relative to every end (that is, every end which does not consist only in preventing the achievement of other ends). These maxims could be used as the *material beginning of justification* (in the sense of the first steps available for every particular practical justification) since they could be stated as reasons in every discussion. In addition to such a beginning of justification, of course, rules are required which regulate the transition from acceptance of such maxims as reasons or – what is equivalent – as norms to a decision on the maxims which are in conflict one with the other in the actual situation. These rules must regulate this transition in a finite number of steps of justification.

I can here only sketch out some possible lines of argument. At least those maxims can be stated as a material beginning of justification whose compliance with which is a condition of human life and therefore also of the establishment of ends. It is here a matter of indifference whether these conditions have arisen or been stated only in the course of long historical development or whether they are or seem to be grounded in the "nature" of man. In any case there are also conditions that are determined by the environment and the state of knowledge of the actors and by the ends-culture which is acknowledged as human. These maxims become norms by virtue of the fact that according to the principle of reason it can be justified to everyone that compliance with these maxims should be made possible for all. Starting from such a system of basic norms determined by the conditions of human life, one can then attempt – according to the rational and moral principles – a justification of the particular maxims under discussion. By the steps of rational justification the "material" of the maxims – that is, the ends which should be achieved by compliance with the maxims – is determined. By the step of moral justification these maxims are then to be examined as to whether they are universal or not. What I am here proposing is a reconstruction of the historical development of maxims and norms and, what is more, a reconstruction of this development as a sequence of steps of rational justification which are to be examined with regard to their morality.

Such a rational reconstruction of the development of maxims and norms would satisfy the conditions of application of the principle of reason. It should however be noted that the rules for the application of the principle of reason are not to be understood in the sense that in every conflict situation which arises one has to begin with the search for basic norms and then completely reconstruct the norm differentiation. These rules propose much more a *possibility* of arguing even when the less far-reaching arguments – such as those over the norms which are commonly accepted as superordinate in a certain group – no longer bring about an agreement. To put it more precisely, these rules of application do not state what one *must do in each case* in order to justify a norm or a maxim, but rather what one *can do if* one does not see how to bring about an agreement otherwise.

In such a "state of emergency" with regard to justification the method of rational reconstruction (and the moral critique applied to it) offers a possibility, in acknowledgement of the subject status of those concerned, to achieve an agreement. This subject status, on the one hand, is acknowledged in that the will of those concerned – their maxims and norms – is not decreed by an individual but that this will is reconstructed from within his own development, as the will of independent subjects. On the other hand this subject status is acknowledged in that this development is reconstructed as a rational development like a sequence of arguments and not merely noted and described like a natural process. The application of the practical justification hence at the same time provides the methodical foundation of the historical and social sciences but also compels philosophy to subject itself to questioning as to its applicability. Practical philosophical effort cannot therefore be brought to a conclusion as a completed system of teaching. It remains, necessarily *fragmentary.*

Giving an account of the significance of practical philosophical efforts in their fragmentariness provides us with an opportunity to look back at the development of the above considerations and to attempt an assessment of the whole.

1. The part of the theory of practical justification which is philosophical in the narrow sense – that is, the methodical construction of principles of justification and their pragmatic justification – is in a certain sense merely *formal* in that it neither demands nor generates a knowledge of the will, the opinions and actions of people. It is not, nevertheless, superfluous. The clearest possible formulation of the formal rules of justification enables us to avoid circumlocutory speculation on reason and practice, and permits clear sighted concentration on the material problems of the knowledge necessary for the practical justification.

2. Attempts such as I have made here to construct principles of practical justification with the help of the tools which logic and formal language theory have provided us are sometimes associated with the opinion that they are made in the conviction *that we can forcibly bring about the good life and just acting as a result of the right use of language.* I hope I have succeeded in eliminating this misunderstanding from the very outset. The sense of the use of formal means of argumentation also lies in the fact that the treatment of our material problems, that is, of our problems of life and actions, can be taken up more clearly.

3. The considerations on the theory of practical justification outlined above were at the same time intended to characterize the relationship

between philosophy and the sciences. If the philosopher endeavours to formulate a highest principle of justification – that is, to formulate the Categorical Imperative or the principle of reason – he has not, in doing so, *solved the problem of practical justification*. But he has *clarified the task of practical justification* to an extent that it can also be understood as a task for the sciences. The clarification of the foundations of our knowledge – that is, the clarification of the aims, the (re-) construction of the basic concepts and the setting up of the basic rules for the acquisition of knowledge – is a principle task of the philosopher. It enables him to offer the scientist useful tools for this work. And it compels him to also direct his considerations to the problems of the scientist and to measure himself against the background of meaningful possibilities of applications – the assessment of which no longer lies within the sphere of his competence alone.

NOTE

[1] The present article is the text of a lecture which was intended to outline the conception of "constructive ethics" – from the point view I held in 1975. This conception is outlined in more detail in my *Philosophie der Praxis. Versuch zur Grundlegung einer Lehre vom moralischen Argumentieren in Verbindung mit einer Interpretation der praktischen Philosophie Kants*, Frankfurt (Suhrkamp Verlag) 1971, 2nd ed. 1980.

CARL FRIEDRICH GETHMANN

PROTOETHICS: TOWARDS A FORMAL PRAGMATICS OF JUSTIFICATORY DISCOURSE

1. METHODOLOGICAL CONSIDERATIONS

The traditional understanding of the task of practical philosophy involves the claim that it is possible to make generally acceptable utterances about what human beings should and should not do. Utterances which are prescriptive in this sense, that is, usually directed to specific (individual or collective) addressees are in the first instance common elements of everyday communication. To this extent they are not different from descriptive utterances. Hence, with respect to the rule systems constituting the prescriptive types of utterances there cannot be stated a "descent" of rationality. Someone who says:

I direct you to come here, but remain there,

is breaking the rules of success of linguistic actions in just the same manner as someone who says:

I assert that the ball is red, but it is white.

If one draws the line of demarcation between everyday and scientific utterances at the point at which validity claims are no longer addressed to specific addressees, but rather, in a sense which will have to be specified, have become "universal" validity claims which are addressed in principle to everyone, then there is in the first instance no reason why only descriptive utterances should be found fit to be scientific validity claims and prescriptive utterances put aside as unfit.

On the other hand, *skepticism* concerning the very possibility of practical philosophy is certainly justified if one interprets its program to require that we are to gain a privileged insight regarding what should or should not be done by everyone by means of some particular cognitive effort. Philosophy – in contrast to myth – lays no claim to a privileged source of knowledge, by virtue of which philosophers dispose of insights which are not available to other mortals. Thus, one who is skeptical concerning the very possibility of practical philosophy should first ask him or herself if it is really fair to define the project in terms of

Robert E. Butts and James Robert Brown (eds.), Constructivism and Science, 191–220.

a claim which, in view of the existence of modern science, no one would raise (any more) within the realm of theoretical philosophy. Just as philosophy never claims to make assertions concerning states of affairs which are properly the object of the scientific discourse, it makes no sense to demand it to justify directions the validity of which can only be decided in corresponding discourses.

If one assumes – at least for purposes of argument – that descriptive and prescriptive validity claims can be critically checked against rules in discourses then the genuine task of philosophy consists in making qualified suggestions or recommendations concerning those rules in virtue of which acceptability is evaluated within discourses. Correspondingly I shall take the task of philosophy to be that of constructing suggestions and recommendations for pre-discursive agreements.[1]

A practical philosophy which was developed along these general lines would thus be neither a *normative ethics* (it would not generate any elementary directions) nor a *descriptive metaethics* in the sense of analytically descriptively reporting how people de facto speak in discourses. Rather, by reconstructing the communication rules of prescriptive discourse which in the life-world are always already realized, it yields those building blocks out of which criteria for the critical evaluation of universal directions can be constructed. It is thus based on the success of discourses concerning the "first" and most primitive directions which are posed in the life-world, and is in this sense a *protoethics.* Thus, we can describe as protoethical those rules which are followed in an on-going cultural commerce with directions in everyday life, rules with whose aid it is only possible to decide concerning the correctness or incorrectness of directions raising universal validity claims.

Forms of speech raising validity claims which are always practiced in the life-world (and concerning whose foundation the philosopher as such knows nothing) are called *discourses.* Discourses are of a *descriptive* or *prescriptive* nature depending on whether the validity claims which are evaluated in them are raised in constative or regulative speech acts. Today, almost no one would raise doubts concerning the possibility, at least in certain cases, of conducting descriptive discourses in such a manner that it is possible to single out certain assertions as being *valid for everyone.* And should someone raise doubts, it generally suffices to recall the example of validity claims in the sciences. In contrast, there are many people who consider the analogous possibility for prescriptive discourses to be out of the question. The claim that something is impossible is best refuted by producing an acceptable counterexample. The following reflections thus have the task of demonstrating the thesis that it is quite possible to reconstruct certain types of prescriptive discourses such that we can single out directions which raise a claim to being valid for everyone. Thus, directions (of various kinds) will turn out to be no less rational than assertions. But they by no means obey the same standards of rationality. Prescriptive rationality is a rationality *sui generis*. The *justification* of direction follows other rules than does the *substantiation* of assertions, but it does not follow rules in the sense of universal, purposeful prescriptions for action. Most skeptics with reference to the rationality of prescriptive utterances make the mistake of taking the justification of directions to be strictly analogous to the substantiation of assertions, if not that of reducing the former to the latter.[2] The critique of this possibility, e.g. the critique of the so-called naturalistic fallacy, yields for these critics the impossibility of a rational foundation of directions. With regard to such misunderstandings, the important thing is to reconstruct the specific forms of the redemption of prescriptive validity claims as always functioning in the context of the life-world and, on this basis, to crystalize those rules whose following guarantees the general acceptability of directions.

The process of reconstruction consists of two steps: *first* there is the task of articulating actual discursive actions by means of an appropriate terminology. Here one must demonstrate that the recommended terminological differentiations and the actions which are characterized with their help are means to the end, namely, to the end of dealing with or avoiding conflict among human beings. This kind of terminological labor is a general characteristic of philosophical investigations; many philosophical controversies are of a terminological character, even if they are often mistakenly conducted in the form of descriptive or prescriptive discourse. *Secondly*, using the elaborated terminology, one has to reconstruct and justify the rules, the following of which is a means of attaining the end in view.

Within this realm, terms and rules refer to actions, namely to speech acts in discourses. But this does not mean that we can defer in some sense to an independently existing *theory of action*. On the contrary, to the extent that actions are followings of directions a theory of action is methodically dependent on the kind of reconstruction of discourse which is attempted here. But this does not mean that this approach is presuppositionless in the sense that it would have to produce its own theme, discourse concerning directions. Rather it must be assumed that those forms of action which we have all mastered in the course of everyday life (e.g. elementary directions), along with ends which are *de facto* recognized, are already available in the context of a "culture of discourse" before we, on this basis, can pose the question, e.g. concerning universal directions. Note that the terminological and regulative reconstruction of the elementary actions of the life-world, a reconstruction which is only possible if already in everyday life an at least elementary self-thematization ("reflexivity") of linguistic action is possible, is thus referred to already existent "skills" and is thus more of an art than a science. For this reason, I suggest giving this discipline the title "*pragmatics*".

Human communication and cooperation commonly proceed without disturbances. But if such disturbances should appear, or if they are to be avoided, and if collective and/or individual experience teaches that the redress of disturbances via communicative action is more practical than other forms of redress, such as the violent repression of alternative paths of action, then human discoursive praxis receives a directly instrumental significance for our shared practical and technical mastery of the world. One who wants to successfully participate in this kind of communicative action and in the modes of cooperation which are based on it, must have acquired the elementary skills. By means of these skills it is then possible to open up further possibilities of action in teaching and learning situations. At any rate, the ability to perform directions belongs among the primary capacities with which pragmatics has to deal. Hence, in the pragmatic reconstruction of the act of directing we have chosen an elementary point of departure with reference to which we can assume that the problems which we encounter in this context and the solutions which can be developed are of sufficient general significance. Of course, the "justificatory" recourse to the life-world in the sense of the basic competent actions makes *no fundamentalist claims*. If one should imagine a human culture in which no one is capable of performing directions to others to do something, no philosopher is in a position to prove that such a state of affairs is in principle impossible; on the contrary, we must immediately admit that the problems which we shall attempt to solve here simply would not exist in such a culture. But this is no more evidence against the general validity of these solutions than the fact that only a very few cultures have extricated the mechanical knowledge in the sense of modern physics is evidence against the general validity of this mechanics.

Terminological recommendations with a view toward the expedient reconstruction of human actions are thus not to be understood as analyses or descriptions of linguistic usage. It is rather a question of investigating practiced actions as to their elements, and introducing to this end terms which should, at best, exhibit a certain degree of convergence with *de facto* linguistic usage (which is, of course, itself quite diverse). On the other hand, terminological and regulative reconstruction is subject to another very strict criterion, namely that of *methodical order*. This criterion involves the requirement that in introducing terms we only make use of those terms which have already been introduced. The terminological regress which would appear to develop at this point is in fact not a danger, since we can learn terms not only by relating them to other terms, but also, e.g. by means of deictic actions. There are, by the way, no rules which could determine the compliance with the criterion of methodical order by guiding the individual steps of the reconstruction in a specific direction. Thus, the methodical series cannot lay claim to *exclusivity* as a mark of its validity. The fact that a goal can be reached by taking one path tells us nothing about the availability of other paths. But it is characteristic of the methodical reconstruction, when it is written down, that in it the false turns and dead-ends which have led to infringements of the principle of methodical order no longer make themselves felt. Just as a teacher in the experimental natural sciences, under ordinary circumstances, demonstrates successful experiments before the class, and not the failures, the path which is recommmended here will be presented as one that leads to the goal of a reconstruction of the conditions of the validity of universal directions, a claim that does not exclude the possibility of other paths.

The rules in terms of which the validity claims raised by directions can, under favorable circumstances, be redeemed are of varying status with reference to their degree of acceptability. In everyday life it is a commonplace that what one person finds acceptable is unacceptable to another. This is equally true of directions and assertions. Thus, in everyday life, rules of argumentation in discursive frameworks are in the first place and for the most part situation-variant, or, more precisely, context- and party-variant; we shall call such rules *rhetorical* rules. In contrast, philosophy has traditionally thematized rules which are situation-*in*variant, rules that we shall call *logical* rules. The following discussion will show that logical rules can also be formulated for justificatory discourses.[3] But these are not the only form of "universally binding" rules (as is easily suggested by the common distinction between rhetoric and logic). It will turn out that a class of rules with an intermediary status, namely, rules that are context-variant but party-*in*variant, play a central role in justificatory discourses; we shall call them *topical* rules. Party-invariance is in fact the identifying characteristic for the reconstruction of the conditions of success in redeeming validity claims; in contrast, the additional characteristic of context-invariance is a rather special limiting case. With regard to human discourse, it is important to know whether a direction can raise a claim to being valid for everyone, and less important whether it can do so under all circumstances.

Thus, we are confronted with the task of investigating the question of invariance with regard to the justification of logical and topical rules of discourse. To this end, we must develop pragmatics on a sufficiently abstract level, e.g. doing without singular characterizations of the parties of discourse. It will turn out that the parties need only be characterized as to those actions which they perform in accordance with their "role". A pragmatics which is developed methodically, thematizing invariances with regard to the parties of discourse, will be called a *formal* pragmatics.

2. SCHEMATIZATION

One who wants to participate with some hope of success in teaching/learning situations in which the meaning of terms is learned by carrying out actions must at the very least be capable of performing those actions in the absence of which a learning/teaching situation cannot arise. Directions belong among these "atomic" actions. Understanding directions cannot after all be learned in the normal sense of the term; one who doesn't know how to perceive and/or comply with directions cannot, e.g. be corrected when he is to learn the meaning of the term "hop" by imitating a hopping movement, but thinks that he is in a situation dealing with "watching others hop". A correction of this type is after all only possible with the aid of the type of action which makes a direction (e.g. "imitate this!"). In this sense, the following methodical considerations move in a field of presuppositions: they are addressed to individuals who on the basis of their everyday linguistic praxis already know what directions are (which does not necessarily involve that they are familiar with the term "to direct" or even with the concept "direction").

However, this recourse to everyday linguistic praxis cannot involve the uncritical acceptance of specific validity claims which are raised by directions in everyday life. But this does not hinder insight into the fact that more or less complex validity claims are, in a manner which must be clarified, always co-posited, with directions.

As an initial survey of various types of directions shows, the most simple type is that of "non-end-expressing directions" ("*afinale Aufforderungen*")[4] for which Wittgenstein offers the superb example

(1) Slab!

If one is to understand this direction, one must as it were put oneself into the position of someone who is in a co-operative situation of continuously laying slabs. If the continuity of cooperative labor is disturbed for one reason or another, a specific author directs that a specific addressee hand him another slab. A direction of this type is 'a-final', since the goal of the action which is demanded not only remains implicit (this is also true of the addressee), but is in no sense "open to discussion". Under ordinary cirumstances such a discussion is superfluous, since the goal can generally count as known and accepted on the basis of participation in the cooperative enterprise. After all, most everyday directions are immediately complied with, since they are embedded in cooperative and communicative contexts that commonly proceed without disturbances.

This is possible because of the 'meaning' of actions of this kind. Thus, not merely the meaning of the predicator "slab" but the fact that a direction is being made by the utterance of the word "slab", in the given context is generally recognized. On the other hand, if one speaks *about* discourse in a context which is distant from the situation in question, e.g. in a progress report on the construction of a house, the pragmatic meaning of (1) is by no means clear. One will have at least partially to explicate the situative moments which are implicitly contained in (1) (to what extent it is again a function of the situation in reporting). An example would be:

(2) When the pain forced me to stop working, he directed that I hand him another slab.

This utterance explicitly describes several characteristics of the situation which are necessary for understanding a direction such as (1). The meaning of (1) is, as an explication of this kind shows (if not completely), co-determined by situative conditions. Indeed, in most situations the author of an utterance such as (1) does not want to say that anyone is to hand him the next slab regardless of what the situation might be. (1) is accordingly a *conditional direction*; this terminological recommendation does not prejudice the existence of unconditional directions. In normal speech situations the condition is not, or not completely, explicated, since we can commonly assume that the participants in a communicative situation know just what kind of situation they are in. The less that is the case, the more explicit we must be. Above all in future-oriented situations, in which we by the nature of the case cannot necessarily assume a common understanding of the shared communication situation, we tend to utter conditioned directions with an explicit statement of the relevant condition:

(3) When I give the sign, bring the slab.

Normally, directions of type (1) lead to compliance with the direction on the part of the addressee – but not always. There are two reasons why the addressee might not comply with the direction: *either* he is of the opinion that the situative conditions under which the direction can be complied with are not given (in example (3): he is of the opinion that the sign has not been given), – *or* he is of the opinion that bringing a slab is not the appropriate means for realizing the end in view. In both cases, the addressee – to the extent that he is interested in redressing disturbances to the continuity of work – will direct that the author performs a further utterance. The utterance of the addressee can be called – regardless of its grammatical form – *doubt.* If the doubt concerns the situative conditions, the author of the direction must utter *assertions* concerning the situation; otherwise he will have to make another direction, namely one which refers to the goal of the action. The discursive situation which is produced by such a *goal-oriented direction* ("*finale Aufforderung*") offers sufficient possibilities for a broadening of the terminology we have introduced thus far. With regard to the methodical course of our discussion it should be emphasized that this terminology is neutral with respect to questions concerning the moral ought or the universal validity of imperatives, etc.; no one will consider a discourse concerning the height of a wall to be in any qualified sense a moral discourse. Rather, the concept of the "moral" can be characterized without falling into a vicious circle only with the aid of a detailed reconstruction of justificatory discourses.

This is not the place to continue the discussion of assertions in this direction. Such a discussion would lead to a reconstruction of descriptive types of utterances and ultimately to a constative logic.[5]

Doubt is obviously not the only possibility in reacting to directions. If the addressee speaks to make it known that he is complying with the direction the corresponding utterance can be called *assent.* One can assent to a direction in a strong and in a weak sense. We shall speak of assent in the strong sense if the addressee is in turn prepared to represent the direction in question as author *vis-à-vis* a further addressee (otherwise we shall speak of assent in a weak sense). We cannot investigate the discursive forms of "persuasive speech" which are characterized by weak forms of assent here (the tools of formal pragmatics are not adequate for such an investigation). Thus, in what follows we shall always understand assent to be assent in the strong sense.

Now, the (sad) fact of the matter is that a direction is not always uttered in a context in which, should doubt arise, the addressor is prepared to enter into a discourse concerning the claim he has raised. If, however, this readiness is present (which is operatively immediately obvious should doubt arise), then we shall say that the author of the direction takes over the role of the *proponent* (*P*); the addressee of the demand, who doubts and/or assents is called the *opponent* (*O*). 'Proponent' and 'opponent' are terms for roles which individuals or groups can play and which are univocally characterized by the actions which they involve. If *O* gives voice to doubt concerning a direction, then there is a *conflict* with regard to this direction; if *O* voices assent, then we have a *consensus.* If a consensus concerning a direction is reached, we shall speak of a successful justification of this direction. Discourses which are pursued with the goal of justifying a direction are thus to be called *justificatory discourses*; ('*Rechtfertigungsdiskurse*'); in contrast, discourses concerning assertions are called *substantiatory discourses* ('*Begründungsdiskurse*'). A direction which has been justified relative to specific parties in discourse will be called relatively-justified; if the justification for a direction can be reconstructed for everyone ("party-invariant"), it will be called absolutely justified or *correct.*

The characterization of correct directions requires methodological tools which can make possible an assent concerning directions which are party-invariant. To this end we must begin by laying out those moments which one must minimally know in order to understand the significance of a direction independently of the situation in which it is uttered (thus independently of the context of speech and of the specific parties involved). A person who is capable of successfully uttering a direction is in command of these moments. Thus, we must inquire into the situation-invariant conditions for successfully making directions. A person who disposes of these conditions can, as we shall say, realize the *schema* of a direction. Directions are uttered with a claim to assent. In making a direction one thus presupposes that it is capable of evoking assent, that it is *valid*. A first condition and thus a first moment of the schema of a direction is the presupposition *of a validity claim*. This moment of action will be called the performative moment or *performation* of the direction. Whoever makes a direction further presupposes that something is directed, namely that a state of affairs be realized. This condition will be called the *presupposition of factuality* ('*Sachbezug*'), the corresponding moment of action, propositional moment or *proposition*. Parallel to assertions, factuality can be reconstructed as being univocally determined by the moments of nomination and predication.[6] For explicit communication concerning performation, predication and nomination one requires the corresponding expressions, sc. performative modes ("performators") predicators and nominators, which are available in natural languages or are the subject of explicit conventions. We shall use the sign "!" for the performator of direction. Symbols for predicators and nominators can be established as usual. The explicit standard form of a direction thus takes the following form:

(4) $!_{xy} . F(a).$

(X directs that Y make an F; or that Y give a the property F. n-place predicators are symbolized correspondingly. Propositions can also be symbolized by p, q, . . .)

Thus, the case we have considered up to now, that of a conditioned direction, is expressed by the following formula:

(5) $G(b) \rightarrow !_{xy} . F(a).$

(X directs that Y make an F, if b is G.)

The meaning of "$\rightarrow$" can be elaborated with the help of the performator of assent: the addressee is to make an *F*, on the assumption that he assents to the assertion that *b* is *G*.

For purposes of reconstruction it is expedient to single out cases in which the conditioning situation generally is the case and in which the author wants to make the direction in any case. A demand of this kind will be called a *rule*. It will be symbolized by the symbol "$\Rightarrow$":

(6) $G(b) \underset{xy}{\Rightarrow} F(a)$

(*X* directs that *Y* always make an *F*, if *b* is *G*; we need not symbolize the performator here, since the double arrow, expressing a rule, always expresses directions, whereas the simple arrow can also stand between other performators such as assertion, as subjunctor.)

3. FINALIZATION, GENERALIZATION, UNIVERSALIZATION

The procedure of schematizing indicates how one is to understand the fact that parties in discourses can dispense with directions and make arrangements with each other in a way that is distanced from the situation itself. This opens up the possibility of reconstructing justificatory procedures for directions without going into the propositional content of specific demands. Directions are understood in principle as utterances whose aim consists in overcoming disturbed cooperation. Correspondingly, in what follows the reconstruction of the steps of justification will methodologically be guided by the idea that the possible disturbances in a communicative situation can be described and possible solutions construed, i.e. we shall single out those situations in which the conflict finally yields to a consensus on the basis of which action is possible.

If *O* voices doubt concerning an afinal direction then *P* must make an initial attempt at justifying this direction. Thus, in the context of house-building of an *O* who refuses to continue handing along slabs, a successful justification might be pointing out the state of affairs which is to be realized in handing along slabs. States of affairs which are to be realized by actions are called *ends* ('*Zwecke*'). Thus, *P* could say:

(7) We want to build this wall higher!

Directions of this kind, which explicitly name the end of the action called for by the demand, will be called 'end-expressing' ('*final*') directions ('*finale Aufforderungen*'). The discursive step from an 'afinal' to a 'final' direction will be called *finalization*. In making a final direction, *P* makes a validity claim of a new kind, namely that *P* and *O* are both pursuing the same end. If this assumption is correct and *O* assents to the final direction, then the direction is justified relative to *O*. Whether or not the end which *P* names can be justified thus depends on whether or not a specific *O* is prepared to assent to it. In addition, even a specific *O* will not assent to an attempt at finalization in all contexts. Finalizations are thus both party- and context-variant. Justificatory discourses in the form of finalizations thus follow rules which have a *rhetorical* status.

If it is also the case that the action which *P*'s first direction called for was appropriate to realize the end – if this action is a *means* to the end – then the initial direction too is justified relative to *O*. Whether or not an action is a means to an end is of course the possible object of a discourse; not, however a justificatory discourse, but rather of a substantiatory discourse ('*Begründungsdiskurs*'). Whether or not an action is a means to an end is the proper theme of assertions. In order to substantiate assertions made in answer to questions of this type, one often must acquire an extensive (possibly scientific) knowledge, which is not always available. Thus, instrumental questions of this kind are de facto often tied up with problems of acting under risk or uncertainty; with reference to the readiness to act under risk, a new justificatory discourse is required. One can begin to see just how multiply intertwined factual discourses can be from the point of view of formal-pragmatic reconstruction.

The success of justification of final directions is dependent on the shared ends of the action context. When *P* explicitly states such an end, he assumes this commonality. Commonly, the contextual support for the ends are sufficient in everyday contexts; that is to say, in normal technical and practical context of action, the justification of a final direction will succeed. It need not be assumed that *P* names an arbitrarily disparate end, nor that in the course of normal action one of the participants suddenly refuses to continue to accept the ends which had been shared up to that point. As one can see, the continuation of the reconstruction of justificatory-pragmatic action is oriented in terms of cases which from the everyday point of view are unusually dramatic, but which can indeed come about, cases in which the supposedly common ends do not or no longer exist. On the assumption that the stock of final directions has been exhausted, that is to say, *P* cannot succeed in gaining *O*'s assent by formulating general or specific ends, *P* will have to attempt to comprehend *O*'s deviance from the hitherto (perhaps only supposedly) valid end as a singular act, or, to put it a bit differently: to understand the situation which now exists as a case of a rule which is otherwise accepted. Thus, *P* attempts to repair the commonality of the context of action such that the given refusal can be interpreted as a merely singular occurrence. A corresponding *general direction* might be:

(8) Walls for houses of this kind are always built a bit higher!

With this justificatory appeal to a *generalization*, *P* thus calls on *O* to reflect on the fact that his participation in the practical context manifests a tacit acceptance of a general goal, and reminds him that the conflict can be put aside if *O* recognizes it as an example of this rule. In making use of generalization as a justificatory strategy in everyday contexts, we presuppose the relative constancy of human action. The utterance of the general direction appeals as it were to this constancy. If this assumption were in principle false, the utterance of a general direction would never have a chance of being accepted. But this does not signify that a refusal to accept a general direction must be considered a violation of rules, which must be met with sanctions; it is rather a matter of making use of a proviso to which every generalization is subject (as long as it has not been explicitly institutionalized). Nothing forces *O* to accept a direction which is made in a specific situation if he is determined not to agree. This shows that rules for justificatory discourses with general directions – a large group of the directions which are uttered in educational and training contexts also belong here – , contrary to a very wide-spread assumption, have a merely *rhetorical* status. The fact that something has always been done in a certain way in a situation *S* is only a telling argument for someone who generally agrees to the corresponding imperative, but has failed to note that situation *S* has been realized.

This reference to the rhetorical status of generalizing justifications also points out the inadequacy of generalization principles which rely on the universal quantification of situational features as a sufficient criterion of universalizability. If one were to formulate the corresponding moral principle in the sense of a direction for the constancy of moral judgment, i.e. *require* generalization, the success of the corresponding discourse would still not be guaranteed, since *P* and *O* might accept conflicting general directions.[7]

The justificatory strategy which is used when one utters a general direction is of no little importance with reference to its justificatory capacities on the basis of the, in everyday context, thoroughly "rational" constancy of human action. To continue with the example of house construction, a situation which is generally characterized by constancy: by and large (as a rule), everyone who is engaged in this situation will accept general directions which are normally presupposed as valid for this situation; this is just the way things are done when one is building a house.

But we must consider which possible solutions are available when the disturbance of a cooperation is not redressed by general directions. On the assumption that all possibilities on the level of generalization have been exhausted – which in the concrete case can only be determined by means of a long series of attempts – we must go on to ask how general directions are in turn to be justified. With the failure of the attempt at generalization, *P*'s attempt to treat the existing situation, which was to justify the direction as a case of a rule which is itself acceptable, has failed. In casting doubt on this direction, *O* demonstrates that he considers himself to be in a singular situation. The fact that he has always done something in a certain way no longer counts for him. Thus, if he is to have any chance at all of justifying his directions, *P* must attempt to utter a direction whose claim to validity does not refer to just this doubting *O*, but rather to every (in this situation) possible *O*. A direction of this kind might be:

(9) Everybody has to lend a hand here!

This type of direction, which is distinguished from general directions by the fact that *O* now appears as an element of a class of individuals, can be called the *universal direction*, the corresponding level of justification *universalization*. This step is particularly relevant in those cases in which the parties represent conflicting directions.

Since – as we have assumed – assent was impossible on the lower levels of justification, the parties must now defer to a direction which is not merely relative, but rather absolutely valid, i.e. *correct*, with reference to everyone and thus also to the parties of the conflict. Directions which raise a universal claim to validity are thus characterized by the fact that they universally quantify over the addressees.[8] Universal directions which are assumed to be justifiable, will be called *norms*, whereas afinal, final and general demands will be called *maxims*;[9] *imperatives* are divided completely and disjunctively into norms and maxims.

To the extent that universal demands legitimately raise the claim to being vaild for everyone, they are party-*in*variant. But it is by no means necessary that their validity be free from conditions. At any rate, unconditioned norms are rare in everyday life. Even explicitly unconditioned norms often turn out to be implicitly conditioned. For example, the commandment not to kill is generally given an unconditional formulation, whereas most moral and legal systems allow that it is permissible to kill in, e.g. self-defense. The conditioned character of norms does not, however, affect their universality, since what is important is validity for everyone. Thus, it turns out that the rules of universalization are by and large *topical*. We shall deal with the extrication of logical rules at a later point.

"Norms" in the sense introduced here are prescriptive utterances; they should not be confused with those descriptive utterances with which one asserts the factual acceptance of imperatives of one sort or another on the part of individuals or groups ("assertions of norm").[10] A norm in our sense would be, e.g.

(10) All Frenchmen must drive on the right-hand side of the road!

In contrast, an example for an assertion of a norm would be:

(11) In France one drives on the right-hand side.

It is easy to see that one can formulate utterances which use the same words to express norms and to assert norms. The following might be an example:

(12) In France one must drive on the right-hand side.(!)

Utterances of this kind are thus performatively ambiguous and thus require a performative explication. This *performative ambiguity*, the importance of which can be seen in the fact that assertions require substantiation, whereas directions require justification, is responsible for a series of misunderstandings in the philosophy of science. For example, the fact that assertions of norms appear in the sciences does nothing to change their descriptive character. Sentences of type (11) can appear in scientific context without further ado, though there is no doubt that descriptive utterances can be scientific. On the other hand, this essay has the task of demonstrating that prescriptive utterances by no means take the back seat to descriptive utterances with respect to their "rationality". The manner in which the validity claims which they raise are redeemed is, however, very different. Whether or not one chooses to award prescriptive utterances such as norms the predicate "scientific" thus turns out to be a merely terminological problem. If one can no longer claim that prescriptive utterances are "irrational", it will be difficult to find reasons for the claim that the sciences should contain only descriptive utterances.

Both descriptive and prescriptive utterances are characterized by the fact that the "genealogy" of an utterance raising a validity claim, i.e. the fact that a validity claim is first raised (culture-historically) at a specific point in time, prior to which the corresponding utterance was unknown, does not affect the validity claim itself either affirmatively or destructively. The validity of directions and thus also of norms is decided not by reference to certain properties of propositions (for example, "atemporality") but only by the rejection or acceptance of reasons. If the validity of utterances is approached from the point of view of the pragmatics of discourse, it is not put in question by a *genesis* (indeed, one which can itself be scientifically explained) of the process of making utterances.[11]

Since according to this reconstruction various kinds of maxims – if they can be justified – can be justified by means of norms, the problem of the "correctness" of imperatives can be pointed methodically to the question of the *justifiability of norms*. The question is not whether norms can in fact be justified – this is of course possible by means of the addressee's actions of assent – but rather concerns the strategies which are used or which can be reconstructed as having been used when an assent does not result. In this case – according to the introduction of the concepts "norm" and "justification" – a recourse to that which everyone wills or which can be expected of everyone is necessary. Acquisition of the corresponding knowledge is anything but trivial (in contrast, the problems concerning the justification of maxims can to a certain extent be called trivial, since the justification either follows, or we can defer to justification problems of a higher level).[12] Thus, with respect to the history of science it is not surprising that in this case – as with regard to other aspects of the problem of the universalizability of knowledge and action – an autonomous discipline, that of *ethics*, has been developed. According to the reconstruction of justificatory discourses recommended here, ethics deals with problems concerning the justification of norms, especially with cases in which conflicting norms arise.

Thus, the basic principle of ethics would be the postulate that only correct norms be embodied in directions; Schwemmer calls this principle the "principle of reason", since it expresses the universality of correct directions and thus – to use traditional philosophical terminology – appeals to reason as the faculty of universality. The principle can also take the following form: "Act so that the maxim of your action can simultaneously be valid as a norm."[13]

We shall not pursue the specifically ethical problems here, since that would lead far beyond the realm of the formal pragmatics of justificatory discourse. In this context it is rather necessary that we give a more exact formulation of the methodological problems to which the step of universalization gives rise.

The methodological criterion of correct norms is that of the *party-invariance* of the validity claim which they raise. With respect to the history of philosophy, this amounts to a reconstruction of that universality which is decisive for Kant's "categorical imperative", but without making use of Kant's key mentalistic terms (e.g. "will", "inclination").[14] But the methodological criterion of *party-invariance* has not yet been given a constructive formulation. How can one know whether a norm in fact legitimately claims validity for everyone? The considerations which follow are to be read as contributions to the constructivity of the methodological criterion of party-invariance.

A first step toward clarification can be found in Kant; it consists in restricting the demand for party-invariance to such (possible) parties, who are concerned by a norm. Under those who are concerned we would find, in addition to the authors of the norm, those who are addressed by the norm (but not those who are possibly affected by the results of following a norm, since that would not yield any restriction of the concept of party). At any rate, party-invariance exists only when the addressees of the norm can assent to it; assent in a strong sense means, however, that the addressees must be able to appear in another discourse as author of the norm. In this case, the test for party-invariance would consist once again in possible assent on the part of everyone, whereby now also the original authors must be able to belong to the everyone. In line with these considerations, the criterion of party-invariance is made precise and operationalized to the effect that a norm is correct if the author could also be the addressee of the norm, and the addressee also be the author. One can express this state of affairs by saying that party-invariance involves waiving the distinction (privileging) of specific authors of the norm; it must be fundamentally arbitrary who is author and who addressee of the direction. This significance of universality (party-invariance) of a correct norm is what Kant has in mind, when he speaks of the universality of a legislation and deduces from the categorical imperative "that every rational being must be able to regard himself as an end in himself with reference to all laws to which he may be subject, whatever they may be, and thus as giving universal laws". (Here his legal terminology already metaphorically hints at institutionalized norms).[15]

4. HABITS, TRADITIONS, INSTITUTIONS

With regard to the question whether the addressee of the norm can also be thought as author, the test of party-invariancy already produces clarification of the question of correctness in many cases. Roughly, it excludes privileged authorship of norms as being party-invariant. On the other hand, it is easy to find cases for which this version of the criterion does not yield clarification, since it is impossible to decide whether all norm addressees could be authors, i.e. would be prepared to assent to the direction in the sense that they would accept the validity claim to be a universal direction. This is above all the case when an actual agreement encounters, not technical problems (e.g. in the case of a survey of world opinion), but rather problems in principle (e.g. in the case of decisions which require the agreement of those not yet born, a problem which contemporary discussions concerning ecology has made especially pressing).

The problem of how we are to deal with competence for assent in such cases is by no means a problem which crops up only in a constructivist context. The case of a competence for assent which in fact is not or cannot be exercised is a regular occurrence in everyday life. In many everyday situations the competence of assent is *delegated*, as in the case of education or guardianship. Another type of situation is characterized by the necessity of taking account of or making allowance for *anticipations* of assent. Finally, we assume as a matter of course that forms of (non-institutionalized) *representation* which recognize the competence for assenting are unavoidable. The forms of delegated, anticipated and representative competence for assenting will be called "*fictitious actions of assent*". Fictitious actions of assent are possible because the constancy of human action, in the absence of which social survival would be impossible, generally makes possible the assessment of the assent on the part of others, whereby, as everyone knows, these possibilities can at any moment be challenged by error and misuse. Thus, it cannot be our goal to demonstrate that fictitious assents are in principle successful justifications; it is rather a matter of reconstructing the possibilities of operationalizing the criterion of party-invariancy which is based on the existence of fictitious assents such that correct and incorrect norms are distinguishable in this context as well. In other words, those social realities which are based on fictitious assent are to be reconstructed in such a manner that the conditions of their admission to discourse become visible.

The foundation of fictitious assent and of the forms of socially constant discourse which it constitutes, is the capacity of individuals to "do the same thing again", i.e. to make use of *action-schemata*. Coming to terms with one another concerning schemata for making directions, as reconstructed in Section 1, presupposes the capacity to make use of schemata for acting. We can only say that a person has carried out a certain action if this person can act according to the corresponding schema (a person who can do a cartwheel does something different in doing so than does a person who accidentally "turns a cartwheel" while falling from a roof).[16] If a specific individual really does always perform a certain action under certain circumstances, we shall call this a *habitual way of acting* (*habitus*). Discursive actions such as doubting or assenting can also be habituated. If individuals implicitly or explicitly give us to understand that they always respond to certain directions with doubt or assent, or if one has other reasons for believing (e.g. on the basis of the logical structure of priorities among ends) that one can count on being confronted with discursive modes of action, this will be called a *habituation*[17] of discourse. The habituation of individual action allows other individuals to take into account the resulting constancy of action. If individuals assume reciprocally habituated actions, a type of action arises (*typification*); if a multiplicity of individuals act habitually or typically in common, and if this action is stable over a period of time, thus constituting a group, this habituated or typified action will be called a *practice* (tradition). Correspondingly, shared habituated or typified discursive action leads to the traditionalizing of discourses.

Directions relate actions implicitly (afinal directions) or explicitly (final directions) to ends. This is also the case when the corresponding directions no longer need be actually uttered, since they have become imbedded in habits or traditions. Ends which, on the basis of traditionalized discursive assent, are determined in advance as already (relative to specific discourses: prediscursively) recognized, will be called *values*. This terminological recommendation relates values strictly to ends pursued in actions, to ends which are de facto recognized on the basis of the social constitution. Values are thus by no means objects which are significant as means for reaching ends (*goods*); but they are also by no means prescriptions in terms of which actions can be socially evaluated (*standards*). Finally, values are distinguished from norms in that the latter are directions and not ends; they are directions which are uttered with the claim to correctness independently from the question of actual social recognition. Values are – according to this suggestion – by definition dependent on a specific constitution by means of recognition (traditionalizing).

This linguistic usage can be elaborated by means of an example: The criterion for party-invariancy was (section 3) the reciprocity between the author and addressee of a norm. Let us assume that in a certain group traditionalization has led to acceptance of the demand that only those directions be accepted as correct whose addressee could accept these directions in the strong sense, i.e. could also be the author of these directions. This direction (of a higher order) would have the end of allowing only party-invariant directions to be accepted as being correct. The direction refers the discursive actions of its addressee to a value; we can say that an individual who participates in this value, who thus need only follow those directions whose author he could be, has "*dignity*". Dignity is thus an example of a value in the sense of our terminology, and is thus anything but a metaphysical something or other in the individual, but rather a discursively specific and thus methodologically reconstructable end of discursive action. This terminology moves in a Kantian direction. As Kant writes:

> Reason, therefore, relates every maxim of the will as giving universal laws to every other will and also to every action toward itself; it does so not for the sake of any other practical motive or future advantage but rather from the idea of the dignity of a rational being who obeys no law except that which he himself also gives.[18]

Like all forms of discursive constancy investigated here, traditionalizations have the sense of making fictitious assent possible. In the case of typified, thus reciprocally imputed actions, the achievements of the traditions extend as far as the imputation of the existence of a tradition holds. It is obvious that the parties of a discourse have an interest in supervising the extent to which typical traditions are followed, with an eye to securing the possibility of constant and reliable action. There are many forms of such supervision and control, forms which can be continually broadened by the imagination; but in the final analysis they consist in making use of social identity as a reward. If a tradition is protected in this sense by means of rewards (or sanctions), we shall speak of an *institution* (in a broad sense). *Institutionalizations* are thus traditional, typical actions of assent which are relatively protected by means of social controls. Finally if this protection of an institution occurs by means of (perhaps itself already institutionalized) violence, this will be called an *organization* (institution in the narrow sense). Traditions, institutions and organizations are forms of social reality which can be reconstructed as the realizations of fictitious assent.

Correspondingly, their correctness can be checked by following up those discursive actions which lead to their development (or which must be considered to be responsible for their development). The important thing here is the consideration that tradition, institution and organization, while not dependent on actual actions of assent (this would destroy their effect of unburdening action), are not *eo ipso* divorced from discourse and can thus at any point be made the theme of discourse. The justification of an organization consists trivially in two parts: the justification of the instrumental adequacy of the protective measures of the institutions implied by the organization, and the justification of the institution itself. The justification of the institution leads back to the social instruments of control and the tradition which they control. The justification of traditions thematizes those directions which constitute the action which has become habituated. We have already discussed the manner in which directions can be justified. The recursive procedure which we encounter here shows that fictitious justifications by means of fictitious assent by no means undercut possible factual agreement. Traditions, institutions and organizations are thus not "irrational" structures simply because they are not *de facto* problematized; they would become irrational and would therefore *a fortiori* have to be rejected as incorrect if they should be declared to be in principle inaccessible for discourse.

If the task of issuing requests aimed at everyone (issuing norms and at the same time securing these norms by organizational means) has been transfered to organizations, these norms are called *laws* (to distinguish them from commands, as discussed in Section 5). Laws are not justified by organizations (organizations are legitimated precisely in order to avoid justificatory discourse), but are *put into effect* by them.[19] In conformity with the procedure of terminological construction offered here, justification and effectuation (*Inkraftsetzung*) are on completely different pragmatic levels. Effectuated laws, whose institutional security is realized, receive – so to speak – their *enforcement* (*Durchsetzung*).

We have now reached the point in the reconstruction of fictitious actions of assent, whose further development would be the task of a philosophical theory of institutions ("philosophy of the state"), at which a few considerations concerning the relation between *consent and decision*, the so called problem of decisionism, are necessary. According to our reconstruction, successful justifications are in principle tied to acts of assent, thus, if one will, to "social" acts. (A philosophically clarified concept of the "social" would itself have to make use of terms such as "directions" and "assent".) The pragmatic characterization of successful justification pays no attention to mental processes which occur in the individuals who participate in the discourses. But since actions are distinguished from other forms of behavior by their intentionality, there is good sense in saying that all justifications are based on decisions. To this extent there is no reason to distinguish between, e.g. cognitive and social validity. But it makes no sense to speak of decisions in the case of fictitious assent. After all, they are fictitious precisely because, while they could have been based on acts of assent, they in fact were not. It is often impossible to identify individuals who make decisions in the case of traditions, institutions and organizations, and such identifications are not essential to the binding character of the social validity of such traditions (e.g. just who the fathers of the constitution were is irrelevant for its institutional and organizational validity claim). Thus, traditions, institutions and organization are not essentially constituted by decisions. The question whether prescriptive forms of validity arise by means of decisions or not is thus irrelevant for the recognition of the validity claim of both factual as well as fictitious assent.

5. LOGIFICATION, CALCULI

Directions for which it is possible to reconstruct situation-invariant validity by definition also have party-invariant validity and are thus a fortiori correct. Thus, we obtain a further operationalization of the criterion of party-invariance by means of a procedure which yields a discursively univocal characterization of situation-invariant directions by situation-invariant (logical) rules. This reconstructive procedure will be called the *logification* of discourses. Logical regulations can be worked out for both simple imperatives (maxims) and for norms.

The *logic of imperatives* is often held to be trivial, since the calculus developed on this basis is (optically) isomorphic with the logic of constatives. Nothing stands in the way of a methodologically subsequent abstraction in the sense of a general propositional calculus. But it must be demonstrated methodologically that this can also be justified in the case of regulative performatives. A justification of this kind can also be given in analogy to that of constative (free of modal operators) logic.[20]

The methodological sense of logification lies in the characterization of those directions which can be justified with reference to every context and every party of discourse. The point of logification lies in the specification of those directions which are ("logically") correct simply on the basis of their situation-invariancy and thus are not in need of any further discursive treatment. As rules for the production of such correct directions we can make use of those rules which make possible the shortening and simplification of discourses involving directions while maintaining their justificatory function. Linguistic abbreviations (logical operators) will be introduced for the abbreviatory rules; if discourses are conducted by means of these operators, those simplifying rules are expedient which allow us to dispose of such operators in a controlled manner. Thus, the meaning of the logical operators has been completely specified when we have formulated their discourse-pragmatic introduction and elimination rules. There is no need for a special theory which would subsequently provide the meaningless signs with meanings (semantics). Thus, the construction of a regulative logic requires only two methodical steps: the extrication of the first primitive logical rules of argumentation in regulative contexts (proto-logic) and the development of a calculus which makes possible the controlled construction of further rules and logically correct propositions (logic).

In the introduction of the *subjunctor* we defer to the fact that for the participants in a discourse in everyday life a standing use of predicators is already determined or to be determined by rules. The use of predicators by no means serves only or even primarily the goal of making assertions about states of affairs with their help. They are just as suited to making possible discursive assent concerning directions. On the assumption that the meaning of the predicator F is sufficiently precisely determined by corresponding rules, it can not only be used to *assert* that the object a has property F, but also to *direct* that object a be brought to have property F. We shall abbreviate a direction of any arbitrary type (afinal, final, general, universal), corresponding to the schema for demands (Section 2) by

(13) $!.F(a).$

On the assumption that the parties of the discourse have pre-discursively agreed upon the predicator rule $G(a) \Rightarrow F(\mathrm{a})$, P can attempt to justify the direction that a be made F by using the demand that a be made G as an argument. If O assents to the direction $!.G(a).$, then if the predicator rule is in fact valid he will consent to $!.F(a)$. This kind of discursive move will be called (material) *inference*. This kind of inference is clearly context- and party-variant and thus has rhetorical status. We shall use the performator $.q.\!\hookrightarrow .p.$ (*consequence*) for (material) inference. An abbreviation of the justificatory discourse is possible in the following sense. If P directs an action in which a is made to be F, and if he can already see that in the course of the discourse he will have to base this direction on the direction that a be made G, then he can immediately anticipate the intention to attempt to use an inference of this kind as a justification. P thereby announces an act of inference at the opening of the discourse. It can be given the schematic form: $!_p/ .q \rightarrow p$. This has the pragmatic meaning: P directs that p be done and will attempt to justify this with q; or: if O assents to q, P directs that p be done. The validity of this demand is in turn dependent on reasons, of course, which must be investigated in terms of rhetorical rules. But independently of the context and the participants it holds that whenever an inference relation obtains among directions, one can immediately go on to a direction under the condition of the inference. This can be represented schematically in the following manner:

($\rightarrow$I) $.q.\!\hookrightarrow .q. \Rightarrow !.q \rightarrow p.$

This is a rule of abbreviation which simultaneously introduces an operator (schematically abbreviated by "→").

Now, if a direction appears in a discourse with → as the opening direction then the complex direction can once again be simplified by means of correspondingly contrary considerations. This direction is to signify, after all, that the proponent directs that p be done, on the condition that O assents that q is to be done:

$$(\to E) \quad !.q \to p., \Vdash_O .q. \Rightarrow !.p.$$

This is also a rule for the elimination of this operator, which is a means for the representation of a discursive sequence of actions. This operator is called a *subjunctor*, a proposition constructed with its aid a *subjunction*; the expression "if q, then p" is normed by the introduction and elimination rules.

The pragmatic characterization of the *negator* can make use of the rules for the subjunctor as well as that form of direction which is familiar to everyday life which is to bring us to precisely not make an object F. On the assumption that P demands that p is to be done under condition that q, but that p does not meet with assent (without justification; abbreviation $\dashv\!\Vdash$) then the reason for this can only be found in the fact that O rejects the direction to do q and is thus of the opinion that q is not to be done. We have the following as a rule for the introduction of the negator:

$$(\neg I) \quad !.q \to p., \dashv\!\Vdash .p. \Rightarrow !.\neg q.$$

is the schematic abbreviation for the omission of the corresponding action (*repulsion*). If a series of antecedents in a discursive context yield the direction to do p as well as to abstain from p, then there must be an incompatibility among the antecedents; at least a portion of them must be in fact unjustified. Since this holds regardless of the situation, we can speak of logical incorrectness here; we shall use the sign for falsity $\curlywedge$ as our abbreviation. The rule for removing negators is:

$$(\neg E) \quad !.p.,!.\neg p. \Rightarrow \curlywedge$$

The expression "not" in its discursive usage is normed by the negator rules.

The *conjunctor* and the *adjunctor* are given a parallel characterization to constative discourses.

(∧I) $!.p.\ !.q. \Rightarrow !.p \wedge q.$
(∧E) $!.p \wedge q. \rightarrow !.p.;\ !.p \wedge q. \Rightarrow !.q.$
(∨I) $!.p. \Rightarrow !.p \vee q.;\ !.q. \Rightarrow !.p \vee q.$
(∨E) $!.p \vee q.,\ !.p \rightarrow r.,\ !.q \rightarrow r. \Rightarrow !.r.$

The introduction of the conjunctor and the adjunctor is also based on schemata of action which are familiar or to be learned in everyday life, namely the *addition* or *alternative* of actions. The expressions "and" and "or" are normed by these rules.

If one understands the quantors as iterated conjunctor (universal quantification, generalizer) or iterated adjunctor (existential quantification, particularizer), the corresponding introduction and elimination rules can easily be interpreted as infinitively frequent directions to make objects F. The expressions "all" and "some" can be normed for regulative discourse in a similar manner.[21]

If one uses rule-schemata instead of rules, i.e. if one replaces the propositions by propositional variables, the regulations developed thus far already yield a logical calculus of natural deduction which corresponds to that of minimal logic.[22] It can easily be supplemented by an intuitionistic or a classical calculus of regulative logic by varying the negator rules.[23] Once one has a calculus, the concept of (logical) probability can also be formulated for directions. It singles out those directions which are *justified on purely logical grounds alone*, i.e. whose propositions can be formed in the calculus alone, independently of all assumptions. These logically correct directions are by definition of *party-invariant* validity. Regulative logic thereby makes its own contribution to the operationalization of the criterion of party-invariance.

Our discussion up to this point can easily be broadened to include the realm of the logic of modal operators (deontic logic) without any speculation concerning possible or any other worlds.[24] To this end one can make use of the available concept of proof or deduction (abbreviation: "$A < B$" for A is deducible from B; "$<A$" for A is provable) to define the performator of obligation (O). On the assumption that $!.p.$ is logically deducible from a system of directions S, then we shall say that doing p is an obligation with reference to S; schematically:

(DO) $O_S.p. \leftrightharpoons S < !.p.$

Given this concept of the obligation it is possible to define forbidden (F), permission (P) and indifference (I) (always in a non-institutionalized sense!) as follows:

(DF) $F_S.p. \leftrightharpoons O_S.\neg p.$
(DP) $P_S.p. \leftrightharpoons \neg O_S.\neg p.$
(DI) $I_S.p. \leftrightharpoons \neg O_S.p. \wedge \neg O_S.\neg p.$

It is clear that the question concerning that which is obliged etc., is in turn dependent on the validity of S, something which possibly is in turn to be decided with the aid of rhetorical rules of discourse. One obtains logical rules, on the other hand, when one asks which proposition is obligatory etc., with respect to any arbitrary S, i.e. system-invariant.

In formulating the corresponding introduction and elimination rules for O (and in view of the possibility of eliminating the remaining deontic operators via definition, this is sufficient), one makes use of the fact that in any case those directions are obligatory which are already logically provable (since a logical calculus is already available for deciding this question). If, namely, $!.p.$ is logically provable, it is not dependent upon any assumptions at all and thus not upon any given S. The logical significance of the obligation is thus determined by the following introduction and elimination rules:

(OI) $!.p. \Rightarrow O.p.$
(whereby $!.p.$ is only dependent upon those formulas whose main operator is O)
(OE) $O.p. \Rightarrow !.p.$

A notation for the rule-schemata using a tree shape is more practical for working with this natural calculus of deontic logic, which is protologically distinguished from the standard calculi above by the fact that it is formulated in terms of propositions of directions and not by assertions of norms;[25] since the propositions which are used are all propositions of directions the performative can be eliminated. But this does not change the pragmatic significance of the propositions-schema. Only the material consequence operator requires special treatment: the proposition which calls for assent is treated as an assumption (marked by square brackets), which is then to be discharged with the aid of the deduction rules formulated in the calculus (provable propositions are of course free of assumptions). Assumptions are numbered as they appear, their discharging receiving the corresponding index number.

That what is obligatory is also permitted can be proved as follows:

$$\dfrac{\dfrac{\dfrac{\dfrac{[O.A.]^1}{A}\,(OE) \qquad \dfrac{[O.\neg A.]^2}{\neg A}\,(OE)}{\curlywedge}\,(\neg E)}{\neg O.\neg A.}\,(\neg I)^2}{O.A. \rightarrow \neg O.\neg A.}\,(\rightarrow I)^1$$

NOTES

[1] This reconstruction of philosophical argumentation is worked out in C. F. Gethmann 1978a.

[2] The terms "substantiation" (*Begründung*) and "justification" (*Rechtfertigung*) are used – as we shall see – without any fundamentalistic connotations. They are thus not vulnerable to the Münchhausen trilemma; cf. C. F. Gethmann and R. Hegselmann 1977a (which does not yet take the terminological distinction between substantiation and justification into account).

[3] With respect to substantiatory discourse, this is shown in C. F. Gethmann 1979, where the terminological distinction between rhetorical, topical and logical rules is introduced (pp. 57–67).

[4] The methodical construction of imperatives follows O. Schwemmer 1978, the remaining pragmatic terminology – to the extent that it is the same for substantiatory and justificatory discourses – C. F. Gethmann 1979. The terminological conception developed here was already used in the discussion to be found in C. F. Gethmann 1978 and C. F. Gethmann 1979b.

[5] The regulative parallels which were left aside in C. F. Gethmann 1979 (cf. p. 141f.) are thus carried through here.

[6] Cf. C. F. Gethmann 1979, pp. 81f.

[7] This is also true of, among others, Hare's "logical universalizability" thesis; cf. R. Hegselmann 1979, pp. 105–126.

[8] Cf. C. F. Gethmann and R. Hegselmann 1977.

[9] This is a slight modification of O. Schwemmer 1978 and C. F. Gethmann 1978.

[10] This distinction did not receive adequate attention in C. F. Gethmann 1979a.

[11] For more details, cf. C. F. Gethmann 1979b.

[12] For a critique of H. Lübbe's triviality thesis, cf. C. F. Gethmann 1978.

[13] O. Schwemmer 1978, p. 147.

[14] For the reconstruction of Kant's practical philosophy, cf. O. Schwemmer 1971.

[15] I. Kant 1969, p. 64.

[16] The distinction between action and action schema corresponds to the distinction between type and token which is generally made in the theory of action; cf. F. Kambartel 1978.

[17] Some aspects of the terminology which is used in the following agrees with that of P. L. Berger and T. Luckmann 1970, which was developed in line with A. Schütz' conception of "society as objective reality"; the basic idea of a "constitution" of social reality is related to the type of reconstruction pursued here. However, the terminological

suggestions made here have been completely cleansed of all mentalistic/biological connotations, and embedded in the context of the pragmatics of discourse. A more extended discussion of the basic conception of a "phenomenology after the linguistic turn" must wait for another occasion. Important work in this direction can be found in W. Kamlah 1972 (esp. pp. 82–90).

[18] Kant 1969, p. 60.

[19] Lübbe's triviality thesis (Lübbe 1978) is based on a confusion; cf. C. F. Gethmann 1978.

[20] Cf. C. F. Gethmann 1979; in light of the analogy it is sufficient if we merely sketch the justificatory procedure up to the specifically regulative-logical modifications.

[21] The usual restrictions for variables would have to be formulated for ($\wedge$I) and ($\vee$E).

[22] I. Johannsson 1936.

[23] This context is presented for constative logic in C. F. Gethmann 1981.

[24] Cf. P. Lorenzen 1969; P. Lorenzen and O. Schwemmer 1975; R. Inhetveen 1977.

[25] Cf. e.g. F. v. Kutschera 1973. Note that we find only propositions of prescriptive performatives in this calculus, such that the calculus is analogous to S4, but without difficulties concerning normative and naturalistic fallacies.

BIBLIOGRAPHY

Berger, P. L. and Luckmann, T.: *Die gesellschaftliche Konstruktion der Wirklichkeit. Eine Theorie der Wissenssoziologie*, Frankfurt a.M.: Suhrkamp 1970.

Gethmann, C. F.: 'Rechtfertigung und Inkraftsetzung von Normen', in Oelmüller, W. (Ed.), *Normenbegründung – Normendurchsetzung*; Paderborn: Schöningh 1978, 140–153.

Gethmann, C. F.: 'Ist Philosophie als Institution nötig?', in Lübbe, H. (Ed.), *Wozu Philosophie?*, Berlin-New York: de Gruyter 1978a, 287–312.

Gethmann, C. F.: *Protologik. Untersuchungen zur formalen Pragmatik von Begründungsdiskursen*. Frankfurt a.M.: Suhrkamp 1979.

Gethmann, C. F.: 'Zur formalen Pragmatik der Normenbegründung', in Mittelstraß, J. (Ed.), *Methodenprobleme der Wissenschaft vom gesellschaftlichen Handeln*; Frankfurt a.M.: Suhrkamp 1979a, 46–76.

Gethmann, C. F.: 'Genesis und Geltung von Normen', in Oelmüller, W. (Ed.), *Normen und Geschichte*; Paderborn: Schöningh 1979b, 17–23.

Gethmann, C. F.: 'Zur formalen Pragmatik des Negators', in *Philosophica* **35** (1985) 39–67.

Gethmann, C. F. and Hegselmann, R.: 'Deontische Logik und Verallgemeinerbarkeit', in Patzig, G., Scheibe, E., and Wieland, W., eds., *Logik – Ethik – Theorie der Geisteswissenschaften* (Akten des 11. Deutschen Kongresses für Philosophie. Göttingen 5.–9.10.1975); Hamburg: Meiner 1977, 357–363.

Gethmann, C. F. and Hegselmann, R.: 'Das Problem der Begründung zwischen Dezisionismus und Fundamentalismus', *Zeitschrift für allgemeine Wissenschaftstheorie* **8** (1977a), 342–368.

Hegselmann, R.: *Normativität und Rationalität. Zum Problem praktischer Vernunft in der Analytischen Philosophie*. Frankfurt a.M.: Campus 1979.

Inhetveen, R.: 'Die konstruktive Interpretation der modallogischen Semantik', in Conte,

A., Hilpinen, R., and Wright, G. H. v., eds., *Deontische Logik und Semantik*; Wiesbaden: Athenaior 1977, 89–100.

Johansson, I.: 'Der Minimalkalkül, ein reduzierter intuitionistischer Formalismus', *Compositio Mathematica* **4** (1936), 119–136.

Kambartel, F.: 'Symbolische Handlungen. Überlegungen zu den Grundlagen einer pragmatischen Theorie der Sprache', in Mittelstraß, J. and Riedel, M., eds., *Vernünftiges Denken. Studien zur praktischen Philosophie und Wissenschaftstheorie*; Berlin-New York: de Gruyter 1978, 3–22.

Kamlah, W.: *Philosophische Anthropologie. Sprachkritische Grundlegung und Ethik.* Mannheim-Wien-Zürich: Bibliographisches Institut 1972.

Kant, I.: *Foundations of the Metaphysics of Morals* (Transl. Lewis White Beck), Indianapolis 1969.

Kutschera, F. v.: *Einführung in die Logik der Normen, Werte und Entscheidungen.* Freiburg i.Br.: Alber 1973.

Lorenzen, P.: *Normative Logic and Ethics.* Mannheim-Zürich: Bibliographisches Institut 1969.

Lorenzen, P. and Schwemmer, O.: *Konstruktive Logik, Ethik und Wissenschaftstheorie.* Mannheim-Wien-Zürich: Bibliographisches Institut ^{2}1975.

Lübbe, H.: 'Sind Normen methodisch begründbar?', in Oelmüller, W. (Ed.), *Transzendentalphilosophische Normenbegründungen*; Paderborn: Schöningh 1978, 38–49.

Schwemmer, O.: *Philosophie der Praxis. Versuch zur Grundlegung einer Lehre vom moralischen Argumentieren.* Frankfurt a.M.: Suhrkamp 1971 (^{2}1980).

Schwemmer, O.: 'Praktische Vernunft und Normenbegründung: Grundprobleme beim Aufbau einer Theorie praktischer Begründungen', in Mieth, D. and Compagnoni, F. (Eds.), *Ethik im Kontext des Glaubens. Probleme – Grundsätze – Methoden*; Freiburg i.Br.: Herder 1978, 138–156.

JÜRGEN MITTELSTRASS

INTERESTS*

> An interest is that by virtue of which reason becomes practical – that is, becomes a cause determining the will.
>
> I. Kant

1. THE TROUBLE WITH INTERESTS

Interests are basic social phenomena. What we do as individuals, and what we do as societies, is largely a function of what our interests are, or what we take them to be. Thus one might expect the concept of an interest to play a fairly crucial role in the conceptual structure of theories within the social sciences. Surprisingly enough, however, it does not. Take down any dictionary in the social sciences and you are hardly likely to find any conceptual entry under 'interest' (with the exception of compound terms like 'public interest'[1] or 'interest groups'[2]).[3] And most philosophers of science continue to ignore the topic.

How can we explain this striking oversight? One possible explanation is that interests make themselves known so easily that there is little to talk about. People *have* interests in much the same way as people have fears, one simply *comes across* other people's interests, one *carries through* with one's interests, or one *abandons* one's interests as one would a sinking ship. As Hegel noted without further comment: "Nothing is accomplished without interest".[4] Anything so easily discerned and discussed can hardly involve major difficulties.

And yet, the view that the concept of an interest is an easy one to understand is arguably deeply mistaken. The difficulties involved in understanding the concept begin to surface when we undertake an examination of various proposed definitions of the concept of 'interest'. There are at least three different types of definition involved. First, there is the economic type. It defines an interest as "the price, or rent, paid in money in exchange for the use of a sum of money".[5] This definition is too specific and technical to throw any light on the general use of the term 'interest'. Secondly, there is the type which, following its use in economics, identifies interest with *utility*. For this type two examples: (1) "When a number of men unite for the defense, mainten-

Robert E. Butts and James Robert Brown (eds.), Constructivism and Science, 221–239.

ance or enhancement of any more or less enduring position or advantage which they possess alike or in common, the term interest is applied both to the group so united and to the cause which unites them".[6] (2) "A man's interest is whatever is profitable to him, whatever helps him get what he wants or what will satisfy him."[7] This identification, again, confines the term 'interest' unnecessarily to its economical meaning, making it, at the same time, unfit for a proper explication in a wider sense that is required in the social sciences.

The same is true of the third type of definition which, in a way, is also the most general type. Here, the definition is given in entirely *psychological* terminology. Again, two examples: (1) "*Interest* in its most general sense refers to a subject-object relationship in which a person (or group of persons) persist in attending to and exerting effort toward some object because of its attractiveness."[8] (2) "In general, an interest is an unsatisfied capacity, corresponding to an unrealized condition, and it is a predisposition to such rearrangement as would tend to realize the indicated condition."[9] In both cases psychological terminology is used in order to describe an *empirical disposition* of man which expresses itself in his acts.[10] What is decisive here is the supposed conceptual identity of sentiments and interests. It is this identity which, as it seems, requires (at least permits) similar analyses. This, however, leads occasionally to determinations which almost move in circles or which are tautological in kind, if, for example, interest is defined as "the relation between a person and anything which he believes will satisfy one of his desires. [. . .] The purposeful and voluntary acts of human beings are always in pursuit of interests".[11] Apparently, no distinction is made here between 'to satisfy one's desires' and 'to satisfy one's interests', one phrase being part of the *definiens*, the other of the *definiendum*.

As we noted three paragraphs back, the definitional failure manifest in these examples is likely due to the seeming intuitive clarity of the meaning of the term 'interest' (as if interests grew in man like flowers grow in the field). But there is also another aspect which adds to this unsatisfactory situation, namely, the use of the term 'interest' in a highly overcharged, conceptually almost uncontrollable terminology. This aspect can be found particularly in Marxist literature; it testifies to a methodologically unreflected Hegelian inheritance. For example, consider the Marxist definition of 'interest': Interest "as a philosophical concept stands for material and ideological relations, realized in their occurrence by individual or collective (social) subjects (individuals, social classes, groups), i.e. for these subjects' material-practical and practical-intellectual, active relations to social conditions representing the objective presuppositions and conditions of their activity".[12] A definition like this already presupposes whole theories about social reality and is therefore absolutely unqualified to support those theories in their definitional parts. Statements sometimes included in the definition underline this particular fact. We can illustrate this by using another example: "The interests represent [...] a mediation between the material circumstances and the ideological (intentions, wishes, etc.). They can only be understood within the historico-materialistically conceived subject-object-dialectic."[13] Strictly speaking, nothing is defined here since one is already philosophizing with the help of terminologies that are either undefined or somehow mutually dependent. This does not mean that statements like these have to be false or unintelligible. It is only that there is no way to argue methodologically for or against such statements on the basis of the given terminology.[14]

Now, bearing all this in mind, my intention in the next few pages is to throw some light on this definitional darkness by introducing a logic or *grammar of interest* that could be applied to theories of social reality within the social sciences. In particular, I shall deal with distinctions like subjective and objective interests or true and false interests. I am not going to argue, however, that there are things like 'objective' interests, or how they might be empirically established. I shall rather try to introduce a concept of objective interest and to stabilize its use in such a way that it becomes clear, without any ontological commitment, just what this concept means. First however, some definitional remarks about the concept of an interest in general.

2. AIMS AND INTERESTS

Interests, whether or not they are in a definitional context, are in some way or other concerned with organizing acts. To pursue an interest means, in a preliminary sense, to exercise an influence on acts, to determine the *effects* or *results* of acts. Because this, again, can only be done by acting, it suggests that acts basically are dependent upon interests. The point in question now is exactly this: Can we attribute a clear meaning to such an expression – without assuming any 'natural' disposition of man or already appealing to entire theories of social conditions?

Acts always refer to *situations*, they are qualified to preserve situations or to change them, i.e. to bring about new situations. Acts, therefore, can also be understood as a *means for the production or preservation of situations*. This, of course, is only one of the aspects under which acts can be considered, but it is an aspect that is particularly relevant for conceptually defining an interest. That acts can be understood as a means for the production or preservation of situations, requires an *interpretation* of these acts. Acts are not immediately intelligible; the intelligibility of acts is rather inferred from the assessment of their instrumental character. We *understand* a particular act as a means for the production or preservation of a given situation to which a *situational description* S always applies. (Indeed, because complete descriptions of factual situations are logically impossible, such a situational description does not refer to 'all' properties of an individual situation. Speaking about situations thus always relates to an abstraction. Here situational descriptions that are logically equivalent turn out to be descriptions of the same (abstract) situation. Accordingly, acts serving the production or preservation of a particular situation have to be taken abstractly.)

If an act a which is itself an instance of A, the *action schema* it represents, becomes interpreted as a means for the production or preservation of a situation s falling under the situational description S, it should now be possible to say that, by interpretation, a particular *interest* is put into the act. In other words: instead of saying that we undertand the act a as a means for the production or preservation of a situation s, we should now be able to speak of an interest which a person P or a group of persons G pursues with the act a. This is not another distinction but only an expression which is equivalent to the original one.

With this explanation, which at the same time reconstructs an everyday understanding of interest, I have followed a suggestion made by F. Kambartel in the context of economic theory: "If a person P or a group of persons G takes (has) an interest [...] in a particular situation s, one can, according to what is commonly used, speak of P's or G's interest in s. Eventually one may briefly call s itself P's or G's interest. In this manner of speaking, for instance, if we have an interest in the institutionally guaranteed freedom of thought, the institutionally guaranteed freedom of thought is our interest."[15] Here, according to the dominant orientation towards basic economic circumstances and especially towards the concept of economic scarcity, particularly such situations are at issue which are qualified to make a satisfaction of *needs* possible. One takes (has) an interest in situations that either directly serve the satisfaction of needs or can be understood in turn as a means of realizing a situation relevant for a satisfaction of needs. In this sense one can also speak of a *mediating context* for interests, i.e. of the mediation of those situations directly relevant for a satisfaction of needs.[16] The following remark makes clear that problems with the concept of interest are not just transferred to the concept of needs. In a manner particularly characteristic of his practice, man is forced to produce situations in which he – e.g. by controlling the distribution of scarce resources – can live. "Practice (which is not understood here in opposition to theory, but simply as the sum of intentional actions and activities) is either itself satisfaction of needs or focuses on bringing about situations in which satisfying needs becomes (directly) possible or through which further practice can bring about such situations."[17] It is in this sense that man can be said to take (have) an interest in his (needs satisfying) practice.

This is now the point where it becomes necessary to differentiate between *interests* and *aims*. To begin with, aims, like interests, can be defined by means of *imperatives* designed to produce or to preserve particular situations.[18] Accordingly, aims or interests are rendered by imperatives in the form "$!S$" ("bring about S!"). Here, again, "S" denotes the situational description applying to a particular situation s, the exclamation mark expressing the imperative. Since in most cases it cannot be decided which acts should bring about a particular situation, one has to distinguish between imperatives designed to produce or to preserve particular situations and imperatives designed to adopt particular acts (*action imperatives* "$!A$"). Moreover, because both specifying aims (or interests) and specifying action imperatives in most cases depend upon the existence of particular situations, we are dealing here, as a rule, with *conditional* imperatives: if a situation s_1 exists, of which the situational description S_1 is true, then *transfer* this situation into another situation s_2, of which the situational description S_2 is true (in standard form: $S_1!S_2$); if a situation s_1 exists, of which the situational description S_1 is true, then *act* in such a way that the action schema A is true of your acting (in standard form: $S_1!A$).

Now, imperatives designed to produce or to preserve a particular situation might be distinguished according to the fact (1) that the means essential for its realization, in particular the relevant action schemata, are available and that, according to the state of the respective knowledge of means, the imperative in question can be followed, and (2) that this is not, at any rate not in a strong sense, the case. In this case we are dealing with *framework* (situational) descriptions which show what one should *aim at* (for example: be concerned about more justice in the distribution of goods!), not with those situations for the realization of which *causal strategies* exist (for example: as a member of parliament vote for the participation model of a particular party *xyz*!). My suggestion is to call those imperatives *aims* which, in this sense, refer to causal strategies of realization. Of course, what has been said about interests in general still applies to this kind of imperative, that is to say, aims too can be represented by imperatives designed to produce or to preserve particular situations, yet in such a way that not every imperative of this kind has to be conceived of as the representation of an aim. Conversely, interests need no longer be understood from the start as situations which can be *causally brought about*. This is always the case when acts are exclusively defined by the interpretative coordination of aims and when aims are understood in this respect as the *effects* of acts.[19]

In fact, both may be true of one and the same context of action. By constructing a good sports field, for example, the *aim* of having such a sports field is fulfilled; nevertheless, it still makes good sense to say that the *interest* in having a good sports field which is linked to this aim, *exists* further on. This interest manifests itself by attending to the field and by taking care that nothing essential becomes changed in the general situation (for instance by preventing excessive use). The interest in having a good sports field, in other words, continues, even after the aim expressed in its construction has been fulfilled. Whereas causal strategies as well as the possibility of realizing them are characteristic for aims, in the case of interests such conditions do not hold. Interests like the interest in having a good sports field only cease to exist if they are no longer a part of our needs or a practice which serves these needs, or if it can be guaranteed that a situation after its realization (as a fulfilled aim) will continue to exist forever. For that, however, there will hardly be a suitable example.

Accordingly, we can now classify aims as special cases of interests. We may also say that aims represent *explicit* cases of interests, distinguished from *implicit* cases which are represented both by framework interests and by those which, according to the given example, are maintained even when the respective aims have been fulfilled. That is to say, aims are seen as depending upon interests. An aim is given (1) where a person *P* or a group (of persons) *G* takes an interest in a particular situation or where, through an interpretation, one may infer from the results of acts that a person *P* or a group (of persons) *G* has an interest in a particular situation and (2) where causal strategies towards its realization are applied:

interests

implicit interests | explicit interests (= aims)

Given these conceptual explications and before I try to reconstruct some parts of the terminology or language of interests used in the social sciences, I should like to make a brief remark about the expression *interest of a group (of persons) G*. Such an interest can be understood as the sum (of interests) of all individual interests represented in a particular group (of persons); in this case the individual interest in this group (of persons) would be *universal*. On the other hand, there is the interest which a particular group (of persons) has agreed upon to the detriment of other interests represented by the same group (of persons). In this latter case, (1) acts of the group (of persons) would serve to establish or to organize the establishment of a common interest (which is no longer the sum of all individual interests in this group) and (2) the individual interest in this group (of persons) would be *singular*. If in what follows the expression interest of a group (of persons) is used, it is used in this second sense of *universal interest*. (The first sense might be understood as a special case of the second, because here the universal interest of the group (of persons) is "accidentally" also individually universal.)

3. THE GRAMMAR OF INTERESTS

To begin with, by introducing the terms interest of a person *P* and interest of a group (of persons) *G* the terms *individual* interest and *collective* interest are also defined:

individual interests $=_{df}$ interests (1) expressed by a person *P* or (2) obtained through an interpretation of the acts of a person *P collective interests* $=_{df}$ interests (1) expressed by a group (of persons) *G* or (2) obtained through an interpretation of the acts of a group (of persons) *G*

Because the terms 'individual' and 'collective' as well as the term 'interest' itself already belong to everyday language, we are dealing here, in the strict sense, not with *explicit* definitions but with (standardized) *explications* of a given language. And just as a reminder: the expression "interest expressed by a person *P* or a group (of persons) *G*" is used, according to what has been said before, with respect to the interest a person *P* or a group (of persons) *G* takes in a particular situation.

Let us turn now to a distinction which constitutes one of the central tenets of a language of interests, the distinction between *subjective* and *objective* interests. This distinction is not intended as a reminder that the situations relevant to a person *P*'s or a group (of persons) *G*'s needs also include those which *P* or *G* – for reasons that might lie in other private orientations or in an incomplete knowledge of the person's or the group's own practice – have no knowledge of at all.[20] On the contrary, so called objective interests are very often invoked against so called subjective interests, subjective interests, as one says, blocking objective interests. If this is so, one says, what matters here, then, is that we transform subjective interests into objective interests. Such a view, however, belongs already to the context of *estimating* a particular social practice characterized by the language of interests, not to the context of explaining or defining what interests like subjective or objective interests are. Hence, we have yet to refrain from these far-reaching claims about the relationship between subjective and objective interests, in favour of conceptual issues.

For these issues the aspect of a *justification* of interests now becomes important. Using again the terminology already established, a justification of interests is to be understood as an argument for the production or preservation of a particular situation which, at the same time, also considers the effects of acts put into operation for that purpose (among them situations which emerge as unintended casual effects). Referring to objective interests cannot dispense with such an argument either, even if sometimes the impression is created that knowledge of a more refined sort, i.e. a special knowledge of the situation, facilitates this effort at least to some extent. The issue in question is, of course, the problem of how to justify practical statements for which there exists no unanimously accepted solution at the moment. This, however, is not at all necessary for our purpose. It is enough to point out that interests, if they are not mistakenly interpreted as a kind of "natural impulse" and, therefore, as a part of nature, are also subject to justifications. Within the framework of our previous discussion, claims about the existence of particular situations can already be looked upon as *arguments* (or *reasons*) for or against the articulation of a particular interest or the call for particular acts. In this case, efforts toward justifications consist in stating the conditional imperatives of the kind $S_1!S_2$ and $S_1!A$ which are followed by the articulation of an interest or by the call for particular acts. If such conditional imperatives are generally accepted as justified, the articulation of interests and the acts which follow them, too, may be taken as justified. Where efforts of this kind are lacking, no claim can be laid that the interests of a person P or of a group (of persons) G are justified. In this case the interests are only *de facto* (*faktisch*) effective or valid.

These considerations now allow for a definition of the term *subjective* interest:

subjective interests $=_{\mathrm{df}}$ individual or collective interests for which, besides a statement of their *de facto* existence (*Faktizität*), no justifications are given

Unlike the psychological definitions mentioned before, but also unlike other definitions which already depend upon theories of social reality, this definition merely makes use of a distinction between the articulation of interests which is oriented towards justification and one which is not. Yet, this definition is by no means "theoretically poorer" than other definitions. This will become clear if we look now at the opposite term 'objective', also applied to interests.

The term or concept 'objective interest' is typically used within the framework of Marxist philosophy and social science. In this case, the whole language of interests is adjusted to this concept. To give an example, objective interests are defined as "a particular disposition, or a particular orientation of the (entire) society, of particular classes, groups, sections or individuals towards an end inherent in the objective social development".[21] Here, developments have ends just as man has interests or aims, and societies have dispositions as, within a psychological language of interests, interests emerge from original (or primitive) impulses. Elsewhere objective interests express "the possibility and the necessity of acting in particular ways already present in the historically determined living conditions of social classes, groups and individuals".[22] In this context G. Lukács is also referring to a "category of objective possibility": "By connecting consciousness with the totality of the society, those thoughts, sentiments etc. are recognized which men *would* have in a particular situation in life if they *were completely able to realize* this situation and the interests resulting from it, both with regard to direct action as well as the structure of the whole society (in accordance with these interests); hence those thoughts etc. are commensurate with their objective situation".[23] Again, the dependence of this concept of objective interests, in its definitional aspects, upon preceding theories of social reality is readily apparent.

Besides this theory-dependence common to all definitional approaches of this kind, they refer, in various ways, to social *developments*. So far as some of these developments are discriminated as objective, the interests which refer to them also count as objective. The question, how one can make these discriminations is, again, answered by pointing at preceding theories, at a knowledge of the "concrete totality" of society, to use Lukács' term. Such attempts are always based on Marxist analyses that use the historical practice of satisfying needs and its requisite social organization as a point of departure. Such analyses (1) represent both the societal conditions and the view of their genesis and mechanism characteristic of these conditions as the result of an interaction between historical practice and the (self-)conception of this practice. And they (2) attribute the evolution of these conditions to so called *laws of motion*, i.e. to dominant *tendencies* that "necessarily" assert themselves.[24] According to this view the causal relationship between a historical practice and the understanding of this practice that developed within it must, in principle, be concealed from those engaged in such a practice. This is why an individual cannot understand its "objective" (interest-)situation – neither directly nor with the theoretical means available to him.

Now, in assessing interests, developments cannot be left out; the intelligibility of situations includes (the requirement of) an analysis of their genesis. That is to say: interests are always part of causal relationships (*Wirkungszusammenhänge*), and this being so, developments (or evolutionary aspects) belong to the definitional context of interests. In the definition of objective interests this fact is taken into account:

objective interests $=_{df}$ individual or collective interests for which a justification is given by an act-interpretative analysis of particular developments

Again, this definition, unlike the approaches mentioned before, does not entail any commitment to particular theories about social reality; it particularly does not entail a commitment to a preceding characterization of particular developments as objective. The proposal rather is to speak about objective interests if the justification of interests actually rests upon an analysis of developments, by supplementing claims about the existence of particular situations (which, after all, function as arguments (or reasons) for or against the articulation of particular interests or chosen acting forms) with claims about the genesis of these situations. What is required, of course, is a method that allows one to argue about an actual genesis with some claim to validity, particularly when contrary assertions and interpretations are in question. But in this case as well, no accompanying theory is necessary which requires us to adopt a dominant development determined by quasi-natural conditions, and which makes the objectivity of interests dependent upon the so-called objectivity of such an assumed development. Moreover, this view would mean also reducing, without exception, justified relationships (*Gründezusammenhänge*) to causal relationships.[25]

Our definition of objective interests as interests for which a justification is given by an (act-interpretative) analysis of particular developments, compared with other proposals such as the distinction between latent and manifest interests, has the advantage of being suitable for the purpose of reconstructions. This, for example, holds true for the concept of *class interest* which is often referred to in our context. In other words, statements about objective interests can be interpreted as statements about *classes*, in the course of which the definition of a class has to be established through an interest characterizing it as a whole. (This characteristic interest has to be directed towards the sum of the relations between all acts to be understood, in order to avoid confusing it with mere groups or group interests.)[26] This procedure, unlike other current definitions of classes, implies that one can get along without any recourse to a concept of (social) power which, in this context, raises additional difficulties that cannot alone be remedied by a terminology of interests and which, in most cases, lead immediately to whole theories (class in itself, class for itself etc.).[27] So much for the distinction between subjective and objective interests.

The distinction between *true* and *false* interests also belongs among theories of social reality. This distinction is mostly linked up with the current use of the term 'objective interests'. Its definition provides no particular difficulties:

true interests $=_{df}$ interests of a person P or a group (of persons) G, obtained through an interpretation of the acts of the respective person P or group (of persons) G, which are pursued with expressed subjective interests in such a way that pursuing subjective interests is a means for the realization of (non-explicit) interests which are always connected to them

Not only subjective interests can appear again as examples of true interests, but also objective interests, which in this way acquire validity *indirectly*. After what has been said about objective interests, this does not mean that these interests have to be distinguished from the start as *justified* interests. True interests, in the sense defined, are almost always *real* interests, i.e. interests which are *de facto* pursued, though indirectly and mostly not transparently to those who do not share these interests. The definition of false interests, then, aims at the context of objective interests:

false interests $=_{df}$ interests which are pursued against objective interests

False interests can be traced back either to the absence of knowledge about situations relevant in the context of a satisfaction of needs, or to (open or concealed) manipulation. In the first case we are dealing with incomplete orientations leading to a false assessment of the state of interests, in the second case with interests that can be interpreted as the effects of true interests. Objective interests, which false interests turn against, refer to either dominant or opposing developments ("play of objective interests"). Under the given circumstances a false understanding of a practice develops, which as such may be typical of a particular stage of developmental contexts ("false consciousness").

As far as I can see, there are no compelling grounds to incorporate the distinction between true and false interests in a systematically arranged terminology of interests. This distinction can be expressed without any problem by using the distinctions already established. Discussing this distinction here, therefore, does not mean continuing to develop a grammar of interests but reconstructing a given, though often not very clear, distinction.

Things are different with a distinction that supplements the concept of subjective interest. Once again, for subjective interests a statement of their *de facto* existence (*Faktizität*) was said to be the defining feature. Through the introduction of the concept of an objective interest, this definition has been supplemented from a systematic point of view in so far as the emphasis on developments has removed interests from mere subjectivity by confronting them with their own genesis. According to this terminology, it is possible to speak about objective interests only with regard to dominant developments or developments which are contrary to them. This, however, implies that the concept of an objective interest alone is not sufficient to establish the distinction between mere *de facto* (subjective) interests and justified interests in a methodologically convincing way.

Now, interests which are neither subjective interests, nor coincide from the start with objective interests, I shall call *transsubjective* interests. This concept requires the possibility of a *rational critique of interests* that includes the justifiability of practical statements:

transsubjective interests $=_{\mathrm{df}}$ universal interests which are justified through a rational critique of interests

This definition permits several other distinctions of note solely by virtue of the now-established terminology, i.e. independently of how the question concerning the possibility of a rational critique of interests is answered. Thus, transsubjective interests might also be objective interests. In that case they would be related to justified developments. On the other hand, transsubjective interests might also be opposed to objective interests, in which case the developments linked with objective interests are not justified developments. Here, to work *against* objective interests needs to be regarded as justified (notwithstanding the fact that it is also possible, with regard to individual or collective interests, to step out of a development that is represented by an objective interest).

As to answering the question how rational critique of interests is possible, I will confine myself to a few remarks in connection with the concept of a *conflict of interests*.

Conflicts of interest are of two kinds: (1) Two interests I_1 and I_2 are to each other like position and negation. In this case, I_1 refers to one of the two conditional imperatives $S_1!S_2$ and $S_1!A$, I_2 to the opposite imperatives $S_1!\neg S_2$ and $S_1!\neg A$ (that is to say: if a situation of the kind S_1 exists, work towards the non-existence of S_2; again, if a situation of the kind S_1 exists, act not in such a way that the action schema A holds true for your actions. To put it briefly: if S_1 exists prevent S_2; if S_1 exists do not adopt A). The situational descriptions S_2 and $\neg S_2$ are mutually exclusive and exhaustive. (2) The pursuit of an interest I_1 results in the prevention of an interest I_2 and vice versa. Here, the prevention of I_2 may be due either to the fact that (a) no action schemata are adopted which realize both interest I_1 and I_2, or to the fact that (b) no action schema exists for that matter. In the first case the prevention holds true only *de facto* (with respect to I_1 a strategy is adopted which impedes I_2), while in the second case it holds true *necessarily* (the classic example is the relationship between paid work and capital). In both cases, the *compatibility* of interests is a *symmetrical relationship*. For that reason the fact that I_1 impedes I_2 but I_2 does not impede I_1, cannot occur. At the same time, this presentation yields a reconstruction of what is meant by *antagonist* interests.

Thus, conflicts of interest are defined either by the fact that action schemata are adopted which refer to contradictory situational descriptions, or by the fact that (*de facto* or necessarily) no action schemata are available which permit the realization of all interests which are articulated in a particular context of action. Conflicts of interest, therefore, cannot be solved just by referring to objective interests. This is also true in the event that objective interests rely on dominant developments. What is needed here, if we are to avoid forcibly preventing some interests from being realized, is a method allowing us to transform conflicting interests into non-conflicting ones. Such a method would need to aim at an *alteration* of interests.

Formulating a moral principle[28] is one of the methods that take this into account. According to this conception, conflicting interests are first interpreted as *derived* (or mediated) interests, i.e. as interests which serve to mediate situations for the relevant immediate satisfaction of needs,[29] and secondly are replaced by *equivalent* interests which do not conflict anymore. Interests are called equivalent with respect to interests $I_1, \ldots, I_n$ if (and only if), in the framework which leads to $I_1, \ldots, I_n$, they can replace each other. In this way, the equivalence relation "*interest equivalence*" (as I would like to call it) for derived interests is defined and, thereby, is the substitutivity of interests. Hence, the moral principle runs as follows: A substitution of his mediated interests can be expected from everyone.[30] Unlike a *compromise*, which is accomplished through a mutual restriction of interests while maintaining the original claims, and which is, therefore, accomplished without any critique of interest, a justifiable alteration of interests is constitutive for the concept of a transsubjective interest which, in fact, was defined by the possibility of a rational critique of interest. Obviously, this method is not sufficient to solve conflicts of interest of all kinds (for instance, with regard to a justification of distribution norms in situations of scarcity). But this is not a fundamental objection to the outlined method. For the definition of transsubjective interests based upon a rational critique of interests, it is sufficient that a method be put forward which, even if it can only be partially applied, guarantees the rationality of such a critique as well as the alteration of interests that follow in its wake.

Within the grammar of interests, i.e. within the structure of an interest terminology, the concept of a transsubjective interest thus plays a central role. It belongs to a *normative* context, whereas other concepts of interest can also be used (merely) *descriptively*. If this is so, the concept of a transsubjective interest takes the place within a grammar of interests that the concept of objective interests adopts in other current terminologies, mostly explained with regard to dominant developments. The present terminology of interests, looked at more closely, shows now a critical vagueness. When objective interests, with regard to so called objective developments, have to fulfill the function of an appellate court with respect to (subjective) orientations, the borderline between *de facto* facts (*Faktizität*) and *normative* facts (*Normativität*) gets blurred. This failure has hopefully been reversed by incorporating the concept of transsubjective interest within the proposed grammar of interests.

NOTES

* Parts of this paper, drawn quite freely from an earlier German version, were read at the Center for Philosophy of Science at the University of Pittsburgh on October 18, 1984. I am very grateful to the Center and its Director, Professor Nicholas Rescher, for

the hospitality I enjoyed as a visiting fellow at the Center and for the very helpful discussion of this paper. I am also grateful to my co-fellows at the Center for their generous assistance in philosophical and linguistic matters, particularly to Professor Robert Almeder (Atlanta), with whom I shared not only many philosophical but also running experiences.

[1] Examples in V. Held, *The Public Interest and Individual Interests*, New York and London 1970, and W. Fach, "Begriff und Logik des 'öffentlichen Interesses'", *Archiv für Rechts- und Sozialphilosophie* **60** (1974), pp. 231–264.

[2] See D. B. Truman, *The Governmental Process. Political Interests and Public Opinion*, New York, 1951, 15th ed. 1968, pp. 33ff.

[3] See R. König (Ed.), *Soziologie*, Frankfurt, 2nd ed. 1967 (only under the heading "Social Change" one reads: "It is to be admitted that *social classes* and the economic *interests* they represent still play an important part [. . .]", p. 296); W. Bernsdorf (Ed.), *Wörterbuch der Soziologie*, Stuttgart, 3rd. ed. 1969; D. L. Sills (Ed.), *International Encyclopedia of the Social Sciences*, vols. I–XVII, New York, 1968 (included are articles on the economic concept of an interest [vol. VII, pp. 471–485], "Interest Groups" [vol. VII, pp. 486–492] and "Conflict of Interests" [vol. III, pp. 242–246; under this last entry reference is made to opinions held in the political sciences and to theories, no conceptual clarifications are given]); A. Görlitz (Ed.), *Handlexikon zur Politikwissenschaft*, Munich, 1970; restriction to a purely economic article also in: J. Gould and W. L. Kolb (Eds.), *A Dictionary of the Social Sciences*, New York, 1964, pp. 341ff. – H. Neuendorf has already drawn attention to this remarkable fact (*Der Begriff des Interesses. Eine Studie zu den Gesellschaftstheorien von Hobbes, Smith und Marx*, Frankfurt, 1973, p. 21) without making any conceptual proposals of his own to redress the conceptual failure at this point. His historical analyses, however, are useful.

[4] *System der Philosophie III* (*Die Philosophie des Geistes*), in: G. W. F. Hegel, *Sämtliche Werke* (*Jubiläumsausgabe*), vols. I–XXVI, ed. H. Glockner, Stuttgart, 1927–1940, vol. X, p. 376.

Accordingly, the concept of interest functions as a basic concept both within theories in the social sciences and within metatheoretical investigations into the aims of the social sciences. Aside from all systematic disagreement in principal matters, this is true of so called pluralist theories as well as of Marxist social theory. Pluralism in political science means a plurality of interests that are organizable and potentially conflicting ("as an empirical theory, the pluralist approach maintains that all political decisions and results can be traced back to group interests", W. Scharpf, *Demokratietheorie zwischen Utopie und Anpassung*, Constance, 1970 [Konstanzer Universitätsreden 25], p. 29 – with reference to D. B. Truman, *loc. cit.*, pp. 14ff.). Correspondingly, one speaks about an "interest-theoretical approach", for instance, in connection with the current discussion of "technocracy" (see C. Offe, "Das politische Dilemma der Technokratie", in: C. Offe, *Strukturprobleme des kapitalistischen Staates*, Frankfurt, 1972, p. 107). Marxist analyses, on the other hand, in no way conflict with such an approach. Engels' often quoted statement that "the economic conditions of a given society first present themselves as interests" (F. Engels, "Zur Wohnungsfrage", in: K. Marx and F. Engels, *Werke*, vols. I–XXXIX, Berlin [East], 1956–1968, vol. XVIII, p. 274), could also be made in any "capitalist" theory. In both cases, "classes" are usually defined by referring to common (objective) interests (see R. Dahrendorf, *Konflikt und Freiheit. Auf dem Wege zur Dienstleistungsgesellschaft*, Munich, 1972, p. 49; C. Offe, "Klassenherrschaft und politisches System. Zur Selektivität politischer Institutionen", in: C. Offe, *loc. cit.*, pp. 68ff.;

G. Klaus and M. Buhr [Eds.], *Philosophisches Wörterbuch*, vols. I–II, Leipzig, 8th ed. 1972, vol. I, p. 537 [entry on "Interests"], pp. 567ff. [entry on "Classes"]).

[5] J. Gould and W. L. Kolb, *loc. cit.*, p. 341 (entry on "Interest (Economics)").

[6] R. M. MacIver, entry on "Interests", in: E. R. A. Seligman (Ed.), *Encyclopaedia of the Social Sciences*, vol. VIII, New York, 1932, p. 144.

[7] J. P. Plamenatz, entry on "Interest (Political Science)", in: J. Gould and W. L. Kolb (Eds.), *loc. cit.*, p. 343. Cf. also W. Hirsch-Weber, *Politik als Interessenkonflikt*, Stuttgart, 1969, p. 94 ("Defined purely formally, interest symbolizes as utility a relation between a person, a number of persons, an institution or a concept and a good whose actual or possible existence, whose state or form benefits or harms the bearer of the interest"); W. Fuchs *et al.* (Eds.), *Lexikon zur Soziologie*, Opladen, 1973, p. 312 (entry on "Interest [2]").

[8] K. Lang, entry on "Interest (Social Psychology)", in: J. Gould and W. L. Kolb (Eds.), *loc. cit.*, p. 344.

[9] A. W. Small, *General Sociology. An Exposition of the Main Development in Sociological Theory from Spencer to Ratzenhofer*, Chicago and London, 1905, p. 433.

[10] In the *Lexikon zur Soziologie* a definition of utility (see Note 7) is preceded by a definition of "subjective" interests directly referred to as psychological: "psychologically the direction of a person's 'attention, thought and intentions' [...] towards an object or fact to which a subjective value is ascribed. Dynamic, motivating characteristic which can be labelled an individual 'attitude towards the valuable' [...]" (p. 312, entry on "Interest [1]"). The definition of interest as *sympathy* (*Anteilnahme*) points in the same direction: "the sympathy of one person towards another person, an object or an event originating in basic vital and psychic impulses or needs" (J. Hoffmeister, *Wörterbuch der philosophischen Begriffe*, Hamburg, 2nd ed. 1955, p. 334 (entry on "Interest"). Or: "Interest as sympathy [*Anteilnahme*] should be understood as both attentive perception [...] and the occupation with something [...] or wishing, desiring, demanding [...]" (W. Hirsch-Weber, *loc. cit.*, p. 95). In all cases, the use of a psychological terminology rests upon the supposedly better manageability of so called *noological* or mental terms such as 'to feel', 'to desire', 'to want'. Strictly speaking, however, this is already a mistake. Here, the manifest reliability of everyday language in everyday life deludes scientific languages into false hopes.

[11] H. P. Fairchild (Ed.), *Dictionary of Sociology*, Totowa, 1973, p. 161.

[12] W. Eichhorn I *et al.* (Eds.), *Wörterbuch der marxistisch-leninistischen Soziologie*, Opladen, 1971, p. 228 (entry on "Interest").

[13] G. Klaus and M. Buhr (Eds.), *loc. cit.*, vol. I, p. 536 (entry on "Interest").

[14] From here to dreams of metaphysics (I. Kant) is often not very far: "Interest [...] is a basic category of man's self-realization that is inherent in him and constitutes him as finite freedom" (F. Ulrich, entry on "Interest I", in: *Lexikon der Pädagogik*, vols. I–IV, Freiburg and Basel and Vienna, 1970–1971, vol. II, p. 305); or (with philosophical reminiscence): "Man not only has interest, he 'is' an *inter-esse*, an intermediate-being [...].The heuristic principle of the 'intermediate' applied to thought reveals it to be an inalienably interested intermediate, because thought itself is a dynamic intermediate, a center, namely the mediation between subject and object, the dialectical connection between the immanent and the transcendent" (A. Esser, entry on "Interest" in: H. Krings and H. M. Baumgartner and Ch. Wild [Eds.], *Handbuch philosophischer Grundbegriffe*, vols. I–III, Munich, 1973–1974, vol. II, p. 746).

[15] F. Kambartel, "Bemerkungen zum normativen Fundament der Ökonomie", in:

J. Mittelstrass (Ed.), *Methodologische Probleme einer normativ-kritischen Gesellschaftstheorie*, Frankfurt, 1975, p. 110. Cf. also P. Janich and F. Kambartel and J. Mittelstrass, *Wissenschaftstheorie als Wissenschaftskritik*, Frankfurt, 1974, p. 111.

[16] See F. Kambartel, "Bemerkungen zum normativen Fundament der Ökonomie", *loc. cit.*, p. 111.

[17] F. Kambartel, "Bemerkungen zum normativen Fundament der Ökonomie", *loc. cit.*, pp. 109f.

[18] See P. Janich and F. Kambartel and J. Mittelstrass, *loc. cit.*, pp. 110ff.; also F. Kambartel, "Moralisches Argumentieren. Methodische Analysen zur Ethik", in: F. Kambartel (Ed.), *Praktische Philosophie und konstruktive Wissenschaftstheorie*, Frankfurt, 1974, pp. 61ff. (here, however, in contrast to the explication given below, the concept of interest is still introduced via the concept of aim).

[19] See P. Lorenzen and O. Schwemmer, *Konstruktive Logik, Ethik und Wissenschaftstheorie*, Mannheim, 1973, pp. 110f.; also O. Schwemmer, "Grundlagen einer normativen Ethik", in: F. Kambartel (Ed.), *Praktische Philosophie und konstruktive Wissenschaftstheorie*, p. 80 ("If the request for an act *H* – holding that *A* is an effect of *H* – is connected with the request designed to produce the situation *A*, then this is saying that *A* is the *aim* of the act *H*").

[20] Cf. F. Kambartel, "Bemerkungen zum normativen Fundament der Ökonomie", *loc. cit.*, p. 111.

[21] P. Bollhagen, *Interesse und Gesellschaft*, Berlin 1967, p. 41. Objective interests are determined "by the objective social conditions of men, particularly by the objective social position and disposition of different classes. [...] All objective requirements of social development are refracted through these classes, obtaining in this way a specific meaning which expresses itself in the directedness of social action" (G. Klaus and M. Buhr [Eds.], *loc. cit.*, vol. I, p. 536). The term 'objective' which is used here so extensively is taken along the lines of a rather carelessly construed classical realism in the sense of 'objective existence' ("the object in question or the phenomenon [exists] independent of the individual subject and its consciousness") and 'objective knowledge' ("the subject [mirrors] the objective reality adequately") (G. Klaus and M. Buhr [Eds.], *loc. cit.*, vol. II, p. 802 [entry on "objective"]).

[22] W. Eichhorn I *et al.* (Eds.), *loc. cit.*, p. 228.

[23] G. Lukács, *Geschichte und Klassenbewusstsein. Studie über marxistische Dialektik*, Berlin, 1923, p. 62.

[24] See K. Marx, *Das Kapital. Kritik der politischen Ökonomie I*, Werke [see Note 4], vol. XXIII, p. 12, pp. 15f.

[25] For the distinction between causal relationships (*Wirkungszusammenhänge*) and justified relationships (*Gründezusammenhänge*) see the German version of this article ("Über Interessen", in: J. Mittelstrass [Ed.], *Methodologische Probleme einer normativ-kritischen Gesellschaftstheorie*, Frankfurt, 1975, pp. 144ff.). The idea behind this distinction is to provide the possibility to speak of justified developments without referring at the same time to dominant developments. A development shall be called *justified* if it can be presented as a series of acts (or action schemata) in which each part presupposes the availability of its preceding parts and if the whole series can be understood as the gradual unfolding of a justified practice.

[26] Cf. O. Schwemmer, *Philosophie der Praxis. Versuch zur Grundlegung einer Lehre vom moralischen Argumentieren in Verbindung mit einer Interpretation der praktischen Philosophie Kants*, Frankfurt, 1971, 1980, p. 217.

[27] Cf., for instance, K. Marx, *Das Elend der Philosophie*, *Werke* [see Note 4], vol. IV, pp. 180f.
[28] See O. Schwemmer, *loc. cit.* [see Note 26], pp. 114ff.; also O. Schwemmer, "Grundlagen einer normativen Ethik", *loc. cit.*, pp. 86ff.; P. Janich and F. Kambartel and J. Mittelstrass, *loc. cit.*, pp. 113ff.
[29] See above, p. 225.
[30] Following Kambartel's reformulation (F. Kambartel, "Wie ist praktische Philosophie konstruktiv möglich?", in: F. Kambartel [Ed.], *Praktische Philosophie und konstruktive Wissenschaftstheorie*, p. 15).

FRIEDRICH KAMBARTEL

IS RATIONAL ECONOMICS AS AN EMPIRICAL-QUANTITATIVE SCIENCE POSSIBLE?

The question – Is rational economics as an empirical-quantitative science possible? – appears to be unnecessary because it can apparently be answered only with a short and final "yes". Is it not the case that quantitative and empirical disciplines have been successfully established in the field of economics for some time? And what actually exists must, finally, be possible. Another question – Is metaphysics possible as a science? – a question that philosophers since Kant have struggled with time and again, was first asked – by Kant himself – at a time when no one disputed the fact that metaphysical disciplines had a place at the universities. And, of course, the metaphysics of Kant's time presented itself as a science, even as the methodologically *primary* science. Kant's question, not unlike the more recent critique of metaphysics by the analytic philosophers, was aimed at exposing the reigning illusion of philosophical thought. In the same vein, I would like to take up the following question. Has not the nearly completed reorganization of economics into an empirical-quantitative enterprise subtly bestowed the appearance and standing of exact science upon pseudo-problems?

But, you ask, isn't it plainly absurd to suppose that there is something wrong with economists conceiving of their discipline as an empirical-quantitative one? That empirical-quantitative economical inquiries have successful applications seems to be undeniable evidence that modern economics works. Yet, in the everyday use of the economic calculus certain of its irrational features become increasingly hard to overlook.

Consider the fact that a cost reduction in automobile production which resulted in a higher accident rate would at present count as an increase in productivity. The same would apply if coffins could be more cheaply produced due to fuller utilization of production capacity. Insofar as the corresponding rise in productivity shows up as an increase in the number of cars or coffins produced (and not, for example, as a reduction in the number of working hours) it appears in our macroeconomic bookkeeping as real growth. Another sort of so-called economic rationality, one that afflicts us in everyday life, turns out to

Robert E. Butts and James Robert Brown (eds.), Constructivism and Science, 241–255.

be pseudorationality in an even better disguise. I am referring to the centralization of public services and administration which has been continually advocated in the past few years, in particular, to the combining of public institutions and offices at central locations. Indeed, labor costs to the public treasury *can* be significantly lowered by centralizing administration and rediviđing the work. However, such a centralization usually means that those who use the public services must pay additional costs for travel and communication. Only if the cost reductions due to the rationalizing process outweigh the burdens subsequently borne by the private citizens using the services, is it possible that the centralization can be economically reasonable.

The examples given may show that it isn't obvious that what at present counts as *economically rational* is also *reasonable*. To use a short formula from our current political discussions, meeting the criterion of economic rationality does not necessarily mean improving the *quality of our life*. The growing awareness of economically based absurdities has led to attempts to supplement the category system of classical economics. Among the new developments is the use of so-called social indicators in place of the usual economic-growth related magnitudes as measures of progress. So far, these attempts have remained without noticeable influence on economists' own concepts of classical economic theory and practice. Indeed, the question remains: Does it suffice to supplement and repair the basic concepts involved in the assertions of economics? Here I want to go into the reasons that suggest that more is necessary: namely, a methodological and practical revolution in our ways of thinking. To that end, I shall next show in broad outline, how economics sees its present-day tasks and procedures.

According to what we read in textbooks, economics is essentially concerned with handling *scarcity situations*. The usual means to that end are thought of as production- (or allocation-)oriented on the one hand, and consumption- (or distribution-)oriented on the other. As a rule, scarcity-situations are taken to be situations in which manufactured or naturally available goods are in *short supply* when compared with the demand for those goods. So much for the economists' characterization of their *task*.

Methodologically, the controversy which has not been settled in the other social sciences seems to have died out in economics: the interpretive (*verstehend*-oriented) school is disappearing and a *scientistic* methodological ideal has achieved prominence. By 'scientistic' I mean tending to organize economics along the lines of physics: as an empirical-quantitative discipline. What do the attributes 'empirical' and 'quantitative' mean here? Contemporary economists think of their discipline as an empirical science in that they, like natural scientists, want to support their material claims with observational data. That is, it is assumed that we have access to the empirical foundation of economics independently of *justified decisions* about the *goals* of production and consumption, the activities that economic theory takes as its objects. To put the matter in an Aristotelian way, empirically oriented economics should not, according to its concept of 'experience', depend on the rational understanding of the "*good life*", that is, on answers to normative questions. In this sense, the empirical orientation of present-day economics implies its limitation to a purely descriptive theory.

There are descriptive sciences that are chiefly *classificatory* in character, as in the case of natural history, e.g. botany and zoology. That economics is modeled after physics implies, on the other hand, giving economic statements a mathematical form by using a basic unit of measurement analogous to the length-, time-, mass- and charge-measurements constitutive of mathematical physics. In so doing, economists as good as presuppose that quantitative claims about economically defined assessments of goods correspond methodologically to physical data.

Certainly, only what is possible can be done. But the mere decision to practice economics as though it were like physics is not enough to enable us to do so, regardless of what we think of the idea. Since Kant, and largely because of him, we are at home with the idea that an empirical science such as physics does not just get underway naturally, as if from itself. Rather, as a successful activity, it depends on securing a series of methodologically prior assumptions. Following Lorenzen, we call the foundational considerations that serve to ground these presuppositions *protophysics*. Empirical, quantitative economics cannot reasonably do without a prototheory either. If economists' view of their discipline as a unified science is supposed to correspond to some possible activity, then a *protoeconomics* must deliver the following: the methodical construction of a universal *measure of value* that is intersubjectively both applicable and reproducible.

Let us look into the possibility of a *protoeconomics* that is relevantly similar to theories of physical measurement. To do so, we should first call to mind the methodological situation in protophysics.[1] In the sphere of classical mechanics, the physical and technical practices of measurement and experimentation can be thought of as a scientific transformation of (among other things) the pre-scientific comparing of forms, movements, and so on. For such purposes, materials must be chosen and the production processes of serially manufactured measuring devices so arranged that these devices are interchangeable with respect to those qualities essential for measuring (for example, form qualities). Consider, for instance, the practice of using calibrated rulers as realizations of rigid straight edges to measure length. The standards necessary for this type of production have won world-wide acceptance in the manufacturing procedures of crafts and industry. A manufacturer who does not adhere meticulously to the calibration norms for measuring instruments defeats the intended use of his products and must reckon with sanctions. At the very least, these may take the form of excess inventory.

The attempt at a protoeconomics must, accordingly, take up the pre-scientific practice leading to the essential concerns of economics defined above: treating scarcity-situations by rationally distributing the means of production and the goods to be consumed. In much the same way that technical physics developed from purposeful activities of the *Lebenswelt*, economics can be linked to the pretheoretical practice of estimating costs and utility in the production and exchange of goods. Thus, the prototheory of quantitative economics has the task of transforming everyday judgments about utility and costs into intersubjective and reproducible ones by using suitable measuring processes. The attempt to do this can also be followed historically in efforts at a quantitative economy from the classics of the labor theory of value right up to modern mathematical economic theory. In what follows, I want to show that this endeavour is methodologically limited in certain fundamental ways when reasonable practice-oriented utility and cost concepts are taken as basic. In so doing, I need not discuss in detail the question of a single unit of measurement for economic value, that is, one that provides cost and utility data in the same dimension. That problem drops out *a fortiori* if distinct prototheories (for cost and utility) which are developed along the same lines as physical ones turn out to be impossible.

First, let us methodically reconstruct the *task* of economics and then determine if and how this task can be pursued *in an empirical-quantitative manner*. To that end, I would like to suggest a series of distinctions that I have already set out elsewhere.[2] Here I will be brief.

We begin with the conception of economics as a theoretical means of treating scarcity-situations. The relevant concept of scarcity-situations, here as in textbook statements of the matter, involves talk of needs and goods. Let us make this talk more precise. Let it refer to purposeful actions, what we would like to call *praxis*. Two types of praxis can be distinguished: the satisfaction of needs, and the actions undertaken to bring about situations making the satisfaction of needs possible. Let 'satisfying a need' mean a praxis not undertaken *solely* to make another praxis possible. Satisfying a need then, results in or effects certain *situations* which represent "ends in themselves", or it is itself a *praxis* done "for its own sake". So far, only the term 'satisfying a need' has been introduced. And now we can proceed to talking of a *need* for those actions or states of affairs which occur as ends or ends in themselves of need-satisfying.

As a rule, needs can only be satisfied in "suitable" situations. For example, certain objects or materials must be available. The needs of a person *P*, or a group *G*, are thus reasons for *P* or *G* to maintain or bring about certain need-satisfying situations. I should like to say that *P* or *G* have an *interest* in such a situation *s*; or, more briefly that *s* is an interest of *P* or *G*. Further, needs fit neatly into this definition of interests as *immediate interests*. Then, we can call a situation *s* a *good* for *P* or *G* just in case *s* is an interest of *P* or *G*. I suggest that '*s* is *of utility* to (*useful* to) *P* or *G*', or '*s* has *Gebrauchswert* ("value in use") for *P* or *G*', is synonymous with '*s* is a good for *P* or *G*'. And talk of *production* is just meant in the sense of *making goods* (things and situations).

We have defined goods as the means of satisfying someone's needs. That is, we have defined them relative to needs actually attributed to certain subjects, in short, relative to *subjective* needs. For this reason, if we want to *evaluate* the production of goods as *justified*, we need another distinction. One must be able to discriminate between *reasonable interests* (in the last analysis, *reasonable needs*), and mere subjective orientations to action and production. To require that orientations be reasonable is to require that, under rational conditions, everyone's agreement to them can be achieved, argumentatively, step by step, in this sense they set a justification. Apparently, one can also talk of justified needs and interests in the sense of this requirement. Because there is, in principle, nothing better available (no ethical intuitions, no derivations from axioms e.g.) I shall consider the controversy about value judgments (*Werturteilsstreit* in relation to Weberian issues) to be settled for this sort of nondogmatic talk of justification and reason.[3]

As a rule, goods cannot be made available for the satisfaction of some needs without adversely affecting the satisfaction of others. This is essentially the case because one has to work for goods. In a broad sense, work can mean any action involved in the production of goods. On the other hand, in a narrower sense, work is understood as a difficult or undesirable activity. To capture this idea, let us take *work in the narrow sense* to be all work in the broad sense that infringes on the needs of the workers. Let us use the term 'labor' in this narrow sense and save 'work' for the broad and the undifferentiated sense. There are two ways in which labor originates:

1. that in which work has a direct adverse effect on the needs of workers (due to noise, stress, etc.)
2. that in which work requires time and thereby runs counter to the general interest of having time to satisfy other needs.

When the satisfaction of needs requires labor, it leads to a scarcity-situation. In such a case, the necessary production of goods cannot be effected such that no needs are left unsatisfied. One must forego satisfying some needs, for either one labors or one defers certain needs whose satisfaction makes labor necessary. In situations of this type one speaks of *shortages*, as opposed to situations in which certain needs cannot be satisfied because certain things and devices have not yet been invented. A shortage exists, according to our definition, when the production of goods can proceed in principle but requires the deferring of interests, either by accepting a cut-back in the consumption of goods or a production-related loss in the quality of life, especially working life. According to current economic formulas though, shortages are simply defined as what obtains when the available or producible goods fall short of meeting the demand for them. However, shortages defined in this way can often be reduced to cases in which the interests pursued make labor necessary. This reduction is always possible when the interests under consideration can be pursued with *producible* goods. If the required goods can be produced without labor then no real scarcity-situation exists. Otherwise, the shortages involving producible goods present themselves in such a way that some needs must go unmet. Either one labors or one defers consumption.

On the other hand, there are obviously goods whose supply cannot be increased by production. Consider land, for example, when it is used for certain purposes. That the supply of a good cannot be increased is economically irrelevant when "substitutes", that is, equivalent, augmentable goods are available. Therefore, only irreplaceable, nonincreasable goods present a scarcity problem not reducible to the necessity of labor. I would like to suggest calling these cases problems of economic '*rationing*', as opposed to problems of *budgeting labor (economic problems in the narrow sense)*. Both sorts of problem can be thought of as *economic problems in the general sense*.

Economic problems in the narrow sense, as well as problems of economic rationing, can arise *unjustified*. It is not only when we are pressured to budget labor or irreplaceable goods that some of our actual interests ought to be changed. For example, driving a car might be criticized with an eye to its dangers to life and limb. And these criticisms would not necessarily lead to any scarcity arguments. Accordingly, we might say that an economic problem or a problem of economic rationing can be posed as a *justified* problem, when it cannot be disposed of by a non-economic critique of the interests involved. On the other hand, economic problems are frequently not apparent on the surface of matters: Reflection may lead us to justified interests which were not seen at the beginning. Abundance (assumed in the first judgment) turns out to be an illusion. A special term is therefore needed for those economic scarcity situations that present a *justified problem*, not just a surface illusion. I suggest calling these scarcity situations '*economically relevant problem situations*'.

Now let us define the basic task of rational economics to be *treating economically relevant problem situations*. Accordingly, economics presupposes as basic the notion of a justified need. With respect to this idea, established economics is largely "economistic": that is, established economics indiscriminately handles all scarcity situations, even those that might be eliminated or at least essentially modified by moral and political criticisms. Thus, the critique of private and state capitalism may prevent us from taking the scarcity situations that are effected by those systems as economically relevant.

With these distinctions in mind, let us return to the original question – Is economics possible as an empirical-quantitative science? To treat scarcity situations in the required empirical-quantitative way, we need measures of cost and utility. Many of today's economists define costs as lost alternative utilities (opportunity costs). Anyone taking this line faces the problem of formulating quantitative economic statements in terms of a *single* unit of measurement. If, as above, we relate our talk of the utility of a good to talk of needs, then a *measure* of utility obviously cannot be based on the actual needs of the relevant subjects. Taken in their entirety, these needs vary and differ so much that the intersubjectivity and reproducibility of quantitative statements, if they were constructed on this basis at all, could not be guaranteed. In addition, nothing can be determined about the needs of a subject without an attempt at "understanding" his or her praxis. Thus, quantitative economics would have to engage in dialogue- and action-theoretical hermeneutics or, we might also say, in a critical *verstehende Soziologie*.

Quantitatively disposed economics has tried to find a way out of these awkward problems by putting the *demand-behavior* of economic subjects in place of the needs of human subjects. (Corporations and similar entities also count as economic subjects.) The pseudorationality of many economic considerations stems from the fact that 'need' and 'demand' (*Bedürfnis* and *Nachfrage*) can easily be mixed up. In German the word '*Bedarf*' is elastic enough to capture the meanings of both words. Needs (and interests) should, however, be clearly distinguished from what is called, in contemporary economic usage 'demand' (*Nachfrage*, *Bedarf*). Demand behavior may become uncoupled from needs originally responsible for it, or it may arise independently of needs. For example, in growing industrial societies with the corresponding norms for capital investments, the course of industrial development may, in part, be circularly determined by industrial expansion. In this case, demand (*Nachfrage* as well as *Bedarf*) is organizationally caused, not determined by needs. On the other hand, it is well known that there are also interests that might never show up as a naturally occurring demand (*Bedarf*). These include, in particular, the so-called "public interests" which can be brought to public attention and satisfied only when the economically relevant subjects unite for that purpose. To rely on the "invisible hand" for the summing up of individual demands would make no sense in this case.

It is also well known that basing the quantification of utility on actual demand (*Nachfrage*) leads to valuations of goods (for example, in the theory of marginal utility) which depend on economic individuals' *potential* to demand goods and thus on the distribution of capital and income. And one can hardly hold that the present historical state of capital- and income-distribution is naturally the result of reasonable rules (of such distribution).

For these reasons, the project of objectifying subjective need structures by using demand-behavior eludes our grasp. It is difficult enough to imagine how one might construct time-dependent utility quanta on the basis of actual demand, quanta which were intersubjectively assertible at a given time. Assuming that this were possible, one still could not methodologically connect the corresponding quantitative statements with the need-related task of economics as defined above. Therefore such a demand-oriented measure of utility (which incidentally has not yet been developed because of internal reasons) would not lead to a proto*economics*.

If one chooses to rationally construct economics beginning with justified interests, then the subjectivity of the given individual needs may be overcome. At the same time, however, the foundation of economics can only be conceived of as the result of reasonable deliberations undertaken by society as a whole. In short, rational politics affects economics. With this claim, we must leave behind the scientistic understanding of economics. Thusfar, this branch of the argument, the *quod erat demonstrandum* holds of a negative answer to the title question of this essay.

The pre-Marxist and Marxist attempts to develop a cost measure (like that in mathematical natural science) remain for discussion. It seems reasonable, according to the considerations advanced above, to conceive of the cost of a good as the quantities of labor and irreplaceable goods that go into its production. Here we may set aside the problem of whether and how a suitably weighted amount of labor-time spent could also constitute a uniform measure of irreplaceable goods. Already a discussion of a *measure of labor spent* will show that we shall not get technical, empirical economic theories that way.

There are several *fundamental methodological* reasons for this. In addition, there are difficulties for which the *present norms of organization* of production and consumption are responsible. The fundamental problems on the other hand, can be subdivided into those which specifically arise from a cost concept developed in terms of *labor spent*, and others of a more *general* nature.

Elsewhere,[4] I have pointed out that measuring spent labor leads to a methodological context in which quantitative and normative considerations are inseparable. Here I only want to summarize my argument. By definition, the various sorts of labor are *inhomogeneous* in the sense that they interfere with our needs to a greater or lesser extent. If we want to get a quantitative "measure" of this inhomogeneity, we must interpret and evaluate the work situation with respect to the needs of the workers. What such an assessment must reasonably show in order to be successful is that the labor expenditure per unit of time increases the more the satisfaction of needs is cut back. The established economic cost-assessment procedure doesn't generally take account of this reasonable norm. These evaluations are based on a systematically distorted labor market, which assigns a higher cost to the work of "higher-level" workers with satisfying working conditions by giving them higher salaries. Thus, using this type of quantitative economics, labor expenditures that cost something (namely, the opportunity of satisfying justified needs) can seem *inexpensive*. On the other hand, what counts as *expensive* labor often coincides to some extent with the satisfaction of needs.

The cost of produced goods can however be "determined" in a non-arbitrary way only insofar as the sorts of work in the production context are *justifiably valued*. A norm, whose goal is to do away with this problem can indeed be *formulated*: namely, that working processes be organized in such a way that they can be regarded as homogeneous. But, unlike the situation surrounding the manufacturing of measuring instruments according to ideal norms in physics, this is no technically approachable idea. Rather, the expenditure-equivalencies between sorts of work lie beyond purely technical evaluation. Furthermore, we have to think of the special inhumanity suggested by the concept of a *wholly homogeneous world of work*.

Further difficulties stand in the way of an isolated technical view of the problem of cost measurement *generally*, not only those of a cost concept based on the value of labor. First among them is portioning out the production costs in *joint production*. In complex cases, the production of goods is organized such that a whole series of different goods are produced *jointly*. *Linked production*, a special case of joint production, occurs when one good originates as a practically unavoidable consequence of the production of some other good, as for example, methane as a by-product of coke production.

If one wants to calculate the *cost per unit of goods* in cases of joint production, then the costs of production must be *distributed* across the jointly produced goods. Joint production is of almost universal significance, for even the production of *one* sort of good can, upon closer examination, be regarded as a case of joint production. Take, for example, the train services of German Federal Railways. When the railways examine their traffic services according to budgetary criteria, then they, for example, encounter the problem of distinguishing the costs of long-distance and commuter trains.

Both sorts of traffic "goods" are administered by one organization and depend on the same technical infrastructure and staff. How these so-called common costs are to be distributed across the various traffic services is obviously no idle question, but a normal consideration in management. This distribution problem is, for instance, important whenever the railways consider dropping a particular service. Although analysts of public finances frequently do it these days, it obviously makes no sense to distribute the costs of timetables according to the percentage of pages a train-type requires, or to reckon the cost of using and maintaining the tracks on the basis of tonnage or passenger-kilometers. Should, in such a calculation, the costs per kilometer of commuter trains prove to be so high that one might want to drop the services, the total costs must be borne by the long-distance traffic, with correspondingly higher costs per kilometer. This is, of course, like 'two plus two' in business management. I mention it here only because I want to draw some unusual consequences from it.

The usual solution, namely choosing an optimal business strategy based on prices acceptable to the public, is methodologically unavailable in our context because we would first have to understand demand-oriented, entrepreneurial individual optimization as an organizational part of a rational, need-oriented economic organization. Furthermore, the problem of joint production would then come up again on the macroeconomic level. Therefore, we have no choice but to enter into a material discussion of the interests behind the production of various goods and their justifications.

First, we must determine which interests the production under consideration is supposed to satisfy basically. After that, we can ask further whether the costs of production are defensible, keeping in mind that the means of production employed could also be used to produce other goods. The question whether the railways produce its actual current services economically, leads, in this way, to questions such as whether long-distance traffic is the Railway's primary purpose. If one answers the latter question in the affirmative, because, say, commuters should rely on other means of transportation such as streets and cars, then one can ask what sort of alternative commuter services the long-distance oriented Railways could economically offer car commuters *as a sideline*.

But it would be another matter entirely if the need situation of those working and living in densely populated areas justifies a primary interest in substituting commuter services for automobile traffic. Suppose it were decided that the primary purpose of the train service ought to be to provide commuter services. In that case the consequences for the overall organization of commuter traffic would have to be drawn accordingly, right down to partially hindering commuting by car in the densely populated areas. In particular, it would only be consistent to count any deficits of the Railways' commuter services due to car users leaving their train seats vacant as costs of the automobile traffic.

Otherwise there would have been no normatively relevant decision, just a wait-and-see attitude. To the extent that this wait-and-see attitude rested on justified decisions, corresponding deficits of the Railways could be accounted for as testing costs of traffic-policy planning. It should be obvious by now that especially in joint production, one can count nothing as "economical" unless one's first step was to bring in the results of need-related, normative deliberations. If our economy is to be organized around reasonable needs and justified purposes based upon them, the problem of calculating production-costs in joint production is not reducible to the sort of technical, intersubjectively invariant, standardization problem encountered in physics.

Another way in which every cost measurement depends on politically specifying rational interests emerges when one figures out the negative side effects of producing and consuming certain goods. What are called *Verursacherprinzipien* in German (literally, "causer-principles") prescribe that the costs of avoiding such side effects count as production costs. Let's call *Verursacherprinzipien* "responsibility-principles". To implement such principles, it must obviously be normatively established which needs and interests may not be impaired, at least not beyond a certain limit. In cost-oriented quantitative economics, responsibility principles are obviously frame rules which standardize the cost-assessments *in general*, that is, which determine their "form". On the other hand, the public discussion of these principles often proceeds as though it were a question of singular, material decisions, more or less analogous to a regulation that cars must be equipped with seat belts. Cars might become more expensive as a result of such a decision, for the *definition* of a suitably equipped car would change. General responsibility principles, however, *constitute* the very quantitative procedures by which costs are calculated. The cost changes with the increase of the use-value, the "quality" of a good is also determined in this way. Here it is important to note that responsibility principles constitute a methodological and thereby unavoidable connection between a reasonable material orientation on our need situations and the cost of producing goods.

In addition to these fundamental or methodological connections between calculating costs and forming justified resolutions on the interests that underlie production and consumption, there are, in the present historical situation, some very general *de facto* obstacles in the way of justified cost-measurement procedures to consider. Our present-day norms for organizing production lead to the appearance of the costs of expanding production capacity among the costs of production. This is because the cost of goods includes investment profits. But the questions, how growth can be made to depend on reasonable interests, and how the necessary costs of growth should be met, require a detailed inquiry. The present system boils down to a growth tax paid by private parties as a sort of tax on consumption and use for which there is no political control. Because this tax is not expressly identified as such and because neither the tax-rate nor the type of taxable goods is rationally regulated, one can assume that it creates essential distortions in cost calculations, when seen against a justified system of cost assessment.

A second historical fact that systematically distorts cost data is that today the *consumption* of products is, in part, not organized as *action*. This means that the intentions behind the production of goods can only be partly enforced. We have already touched on an example illustrating this in our discussion of joint production, namely, that providers of public transportation can only pursue to a limited extent purposes that conflict with the practices of would-be passengers (those of motorists, for example). Another simple example will serve to make this point clearer.

A clockmaker receives orders from two different persons. For each of them he is to make one clock. He finishes the clocks and then receives word from one of the customers that he does not want the clock after all. Let us say there was no reason to think the clock was of inferior quality, and that the right of either party to back out of the deal was understood; that is, that the customer was not obligated to take delivery of the clock. It is then *prima facie* unreasonable for the clockmaker to adjust the bill for the sold clock in order to cover the cost of the one he did not sell. However, that is exactly what happens when, as a principle of covering costs, the cost of unused products is added to the price of goods that were consumed as intended.

Obviously, this is not supposed to be a criticism of the standard practice of dealing with the cost of excess production that occurs because of unforeseen events or because of the indefinite marginal shortages characteristic of every sufficiently complex production schedule. Rather, this example is supposed to refer to those who refrain from carrying out or even deciding on the aims of the production process. Among others, the example refers to those who on liberalistic or bureaucratic grounds leave the subsequent use of products open.

Let us return to our main question. What distinguishes economics from physics with respect to the foundations of quantitative considerations? If they are to have a rational foundation, the specific cost and utility measurements of economics must depend on the *aims* and *organization-rules* of production and consumption. These aims and rules must in turn be founded in the rational deliberation of the members of the particular "economic universe", that is, those who interact economically with one another. These norms can regulate not only whether and how goods are produced, but, more importantly, how the goods are distributed. "Material" orientations of this type cannot be set down once and for all like technical guidelines. In protophysics the situation is different, for here one can justify ideal standards for the manufacture and use of physical measuring devices.

Therefore, there can be no "absolute" universal procedures for measuring cost and utility that permit us to examine the goods and the production processes and just "read off" the values, as we might use scales to determine the mass of a piece of cheese.

This means that when it comes to the methodological construction of economics, economic theory and politics cannot be treated as isolated subjects. Rather, in economics, everything that might count as data for an empirical theory depends itself on the totality of purposes which must already have been taken into account when collecting the data. Quantitative economic information can only be as rational as the system of aims that underlies it. Technical purposes, and judgments about their technically efficient pursuit or realization, can indeed be isolated against a rational totality of purposes pursued. *In this sense*, they can count as value-neutral. On the other hand, reasonable, trans-subjective criteria of assessing utility and cost can only have promise in the framework of a reasonable system of interests satisfied by the production of goods. The usual recommendation to understand economic information *ceteris paribus*, reveals this but also disguises it under the veil of a seemingly value-neutral clause, namely disguises that the *rationality* of the *cetera* (not just the absence of natural or unnatural disturbances of our technical intentions, that is, the *nature* of the *cetera*) must be discussed.

NOTES

[1] For a discussion of questions concerning protophysics see G. Böhme (ed.), *Protophysik*, Frankfurt a.M. 1976.

[2] 'Bemerkungen zum normativen Fundament der Ökonomie', in: J. Mittelstraß (ed.), *Methodologische Probleme einer normativ-kritischen Gesellschaftstheorie*, Frankfurt a.M. 1973, pp. 107–125; also in: F. Kambartel, *Theorie und Begründung*, Frankfurt a.M. 1976, pp. 172–190.

[3] Ethical and methodological problems related to this understanding of the concept of reason are discussed in some detail in: F. Kambartel (ed.), *Praktische Philosophie und Konstruktive Wissenschaftstheorie*, Frankfurt a.M. 1974.

[4] 'Bemerkungen zum normativen Fundament der Ökonomie', see note 2.

PETER JANICH

DETERMINATION BY REALITY OR CONSTRUCTION OF REALITY?

The topic of this paper arises in research into developmental psychology in the form of the problem of the relative importance of external conditions and the subjective orientation attainments of individuals. However, it also represents a general philosophical problem, transcending not only questions of the philosophy of science but even problems of epistemology.

These problems are not to be the subject of the present paper. I am concerned with a narrowly delimited methodological question, the answering of which can be claimed to have direct consequences for the empirical work of developmental psychologists. I thereby suggest that it is incontrovertibly the case that 'ecological developmental psychology', which is concerned on the one hand with 'developments' within the individual and on the other with the 'environment', represents, as an empirical sub-discipline of psychology, a historic fact.

Methodological discussion of this historic fact does not mean the philosopher of science leaving his desk in order to compete with empirical researchers. Neither individuals, nor their development of 'the' environment are the topic of methodological debate but solely the procedures of the developmental psychologist – admittedly accompanied by the risk that consideration of the tacit presuppositions of research into developmental psychology is more likely to be conducive to criticism than to metatheoretical confirmation.

In order to avoid unclarified presuppositions with regard to terminology, I should like to begin with the philosophical question: How is ecological developmental psychology possible as an empirical science? However, even an attempt at answering this question would go beyond the scope of this paper so that, relying on the widely held view that *empirical* research cannot dispense with the *collection of data*, I must further narrow the scope of my question to the following: which implicit or explicit *preconditions* for the collection of data for an empirical ecological developmental psychology are methodologically problematical?

In the first part of this paper I shall describe and criticize various possibilities with respect to preliminary methodological decisions con-

Robert E. Butts and James Robert Brown (eds.), Constructivism and Science, 257—269.

cerning the collection of data in order, in the second part, to delineate an alternative which avoids the errors which have come to light, in the hope that my criticism in the first part is not understood as a general methodological condemnation of a sub-discipline of psychology.

1. METHODOLOGICAL POSITIONS AS PRECONDITIONS FOR THE COLLECTION OF DATA

For reasons of presentation I distinguish four fictitious 'positions', which need not necessarily be identified with schools of developmental psychology or individual researchers – these usually represent blends of these positions. With respect to particular preconditions for the collection of data I should like to discuss:

(a) Environmental Causalists
(b) Individual Causalists
(c) Descriptivists
(d) Relativists

On (a) and (b): *Causalism*

According to the traditional empiricistic view of science widely held by psychologists today, it is the task of psychology *to describe* and *explain* what has been observed, that is to say, *human behavior* in the form of collected data. Regardless of the extensive investigations by philosophers of science into the concept of explanation it is generally understood, in accordance with the everyday use of language, as being the task of hypothesizing, in universal if-then statements, a relationship between 'environmental data' and 'individual data' and subjecting it to empirical control procedures. Depending on the respective approach an attempt is made either to show the conclusions viewed as being in need of explanation to be derivable from suitable premises or to show the premises considered to be in need of explanation as derivable from suitable conclusions. In concrete terms this alternative can consist either in 'attributing' to environmental influences patterns of behavior 'observed' or 'given' on the basis of the collection of data or in asking questions about the 'effects' of environmental components which are also 'given' in the form of data on individual patterns of behavior or their development. The following may serve as an example of the above: A growing lack of concentration is observed amongst a group of junior school pupils and causes are sought in the form of environmental data; or a general increase in class size in the junior school is noted as an environmental factor and its effects on pupils' behavior patterns and their development are then investigated.

Environmentally causalistic is the term which may be given to the procedure whereby environmental components collected as data are taken as a starting point, a relational hypothesis in if-then form is formulated and the presence of the hypothetically asserted individual data is empirically tested.

Individually causalistic, on the other hand, is the term which may be given to the procedure whereby individual data form the starting point and the relational hypothesis is tested by means of the environmental data. The difference between these two procedures is not concerned with the direction of logical conclusion which in both cases leads from statements about 'environment' to statements about patterns of behavior, but with the starting point. This is, in the one case, definite structures of the environment and, in the other, structures of patterns of behavior whereby one or other of these areas is regarded as being 'dependently structured'. I shall return later to the question of this dependence. The general problem, that causal relations are never the result of empirical investigations but are always the result of a previous basic methodological decision, cannot be discussed here. However, such familiar expressions as 'causes', 'effects', 'explain', 'attribute' and even 'independent and dependent variables' etc. (as used above), most often occurring in popular discussions of causality, are more often a source of confusion than a sharp methodological instrument. In the context of our discussion, in which we are concerned with preconditions for the collection of data, the if-then form of their hypotheses is sufficient to characterize causalistic positions. Here if-sentences are to be regarded as statements about environmental components, then-sentences as statements about behavioral components.

Data are collected in order to test hypotheses *empirically*. Data can, of course, only be collected at all where prior distinctions have been made – in the terminology usual amongst psychologists; where definitions of variables are available in the form of operationalizations. As far as I can see, there is here a fairly widespread misconception with regard to the conditions which make a proposition an empirical one. Many assumedly empirical propositions prove on closer examination to be mere linguistic conventions. An example from physics, selected for reasons of simplicity, will be used to demonstrate this point.

According to Galileo objects in free fall within the field of gravity of the earth are subject to constant acceleration. This "law of free fall" can, of course, be empirically tested with the aid of watches, rulers and possibly other measuring instruments, that is to say, one can perform an experiment. What is more, the proposition can easily be converted into an if-then one: "if a body falls freely within the field of gravity of the earth then its movement is constantly accelerated". This Galilean law of free fall is, nevertheless, not necessarily an empirical proposition, for this would mean that it would have to be able to fail any test of experience. For what does 'free' fall mean? Insofar as no operational definition of 'free fall' (freedom from friction for example) has been stated then it is precisely the Galilean law which first *defines* free fall. In other words it is a linguistic convention to call constantly accelerated falling movements 'free'. This is empirically irrefutable and may thus not be regarded as empirical knowledge. The example shows that the question of whether an if-then statement is empirically testable depends on whether all the terms occurring in it are operatively defined.

The explicitness of the definition of all terms is, of course, a necessary condition for a proposition to be empirical, more precisely, a necessary condition for a proposition for which there is believed to be an empirical test, as not to be an empirically irrefutable convention. It is not, however, a sufficient condition. It is rather the case that statements in a language in which all terms have been explicitly defined can already be regarded as logically valid on the basis of definitions. In this case, too, a statement is conventionally immunized against empirical refutation and an 'empirical test', it brings to light a quite different kind of knowledge than that intended by the researcher. Admittedly, the definitional validity of an if-then proposition can only be demonstrated within an adequate terminological framework. But it does not therefore follow that consciously tolerated terminological imprecision serves as a guarantee of immunization against refutation. It is rather the case that connotations of definitions of variables – intended but not made explicit by the psychological researcher – prevent the subsequent provision of an explanation which *excludes* the statements in question considered to be empirically valid. Here, again, a simple example from physics.

"Under suitable air pressure conditions water boils at 100°C". This can, of course, be easily 'tested' with a mercury thermometer. Suppose the "test" is successful. What knowledge has been gained? Contrary to what might be naively supposed, no knowledge is gained of water or its "naturally determined" boiling point but merely knowledge of the thermometer. For the "test" merely supplies the information that the thermometer is functioning as it should; the boiling point of water under certain supplementary conditions defines the 100°C mark on the centigrade scale. Contrary to the assumption, therefore, that one has acquired knowledge of the field of investigation (water, boiling) one has merely acquired knowledge of the functioning of the instrument of investigation (the operationalization) in the sense of one's own postulated terminology. It would clearly be of little use here – to conclude the analogy with the psychological example – to leave the terminology sufficiently open, that is to say, to dispense with the definition of the Centigrade scale. The test as described above could never lead to knowledge of the boiling point of water.

As a non-psychologist I can only suppose that it is not the true intention of psychologists to investigate empirically their own terminology, especially since such investigation is superfluous where all is explicit. On the contrary, they may be assumed to wish to learn something about human beings and their development subject to 'environmental conditions'. The causalist cannot, however, be sure of reaching his aim (knowledge of development dependent on environmental factors) even when his psychological terminology is sufficiently explicit to exclude (1) that a supposed hypothesis is a definition for an as yet undefined term and (2) that the supposed hypothesis is already logically valid on the basis of other definitions. The if-then form of causalistic hypotheses conceals the great and not easily recognized danger that, on the basis of certain operationalizations, the researcher may believe that he is attempting other definitions independently but is in fact constructing them in such a way that hypothetical relational assumptions which have become dear to his heart turn out to be true. This is what was termed above, 'the dependent structuring of the field of investigation'.

Once again an example from physics. Let us assume that in a primitive system of measurement the term 'longer than' is operationalized by comparing the lengths of rods. The hypothesis "If bodies are warmed, they expand" can, before anything is known about its truth or falsehood, mislead one into determining the temperature (seemingly independently) by means of a thermometer after operationalization of the expansion by comparison of lengths. A thermometer is, however, nothing more than a pair of bodies which, when warmed, expand in differing degrees. By sufficiently detailed listing of all the individual steps in the definition it can in this simple case admittedly be demonstrated that an empirical test already presupposes the validity of the stated hypothesis and is therefore pointless. In the case of psychology this would not usually be possible owing to the high degree of complexity of the necessary definitions. Hence, patterns of behavior can probably always be classified in such a way that, while depending on existing definitions for components of the environment, they depend on these in a true but only supposedly empirically true if-then statement. Once again no knowledge of individuals or the environment is gained but, at most, knowledge of one's own terminology. It cannot even be determined whether the former or the latter is the case, that is to say, whether one has arrived at a developmental-psychological proposition about individuals or a metaproposition about one's own terminology.

In summary we may therefore say that the causalistic attempt to test conclusions about individual statements based on environmental statements proves to be unworkable. It is, as a rule, not even possible to achieve the degree of terminological clarification which makes it possible to decide whether we are concerned with a proposition of development psychology. That de facto 'empirical investigations' can be conducted is, as the examples from physics show, no objection to this statement. Admittedly, and this position has wide support among psychologists, there is still the possibility of critically dispensing with statements of *causal* relations and concentrating on *correlations* between variables. I call this position "descriptivistic".

On (c): *Descriptivism*

In the further development of my argument I constantly assume that the definitional validity of problematical statements subjected to empirical testing is excluded. The descriptivistic position – which often claims to be modest – consists in '*describing*' the frequencies of the common occurrence of independently operationalized variables – without asserting causality, frequently even without supposing any causality.

However this supposed methodological abstemiousness achieves little as I shall once again try to demonstrate with an example from outside the field of psychology.

Let's assume that we may, on the one hand, characterize screws and nuts by means of such variables as diameter, length, pitch, etc., and, on the other, by such variables as color, material, the shape of the head, etc. It would then be possible to conduct an empirical investigation in particular areas (for example, the screws from a particular hardware store, screws produced in Germany) to ascertain to what extent, for example, the color black correlates with screws with a diameter of five millimeters. The mathematical apparatus of the psychologist working with correlations could be employed and yet such an investigation would be grotesque – here I do not mean the fact that producers and consumers of screws, ("the man in the street") would question the point of such an investigation. (By consensus of scholars this question is currently impermissible with respect to scientific investigations.) It is grotesque by virtue of the fact that the results of the investigation, whatever they may be, can easily be falsified by the production of an amount of other screws. Screws are artifacts produced for certain purposes and precisely for this reason it is absurd to investigate empirically their characteristics instead of asking the manufacturer for what purpose they were produced. The absurdity of the question, intuitively correctly understood by the non-scientist in his question "what for", lies more precisely in investigating such relations between properties (as, for example, color and diameter) which have no place in the complex of motives for screw production or the uses to which screws are put.

To what extent does this example have a parallel in psychology? It may be objected that, with screws, we are undeniably dealing with artificial products but that psychology and a fortiori ecological developmental psychology has, as its object of study, "that which exists", hence the human being, his development and his environment. Admittedly, a misunderstanding brought about by philosophers and philosophers of science in particular has here created an obstacle which prevents us from recognizing a trivial fact. The empiricist doctrine according to which empirical science is concerned with the collection of individual experiences, so-called *data*, (in English: what is given) and then with the systematization of the given by means of theories, conceals the fact that data are, in a decisive sense, just as much artificial products as are screws. Data are created by the empirical researcher and should terminologically be more exactly described as *facta*, 'facts' (in English: what is made) for it is only on the basis of a system of distinctions which the researcher employs as his starting point that 'data' can be collected, or, as a methodological preliminary, variables can be defined. Not one single piece of data is 'given' in a strict sense but is rather the result of a specific question within the framework of the system of distinctions formed by the language in which the question (empirically; the observational sentence) is formulated. The unavoidable *linguistic* fixation of scientific findings depend on normative presuppositions which cannot be replaced by experience.

The empiricist understanding of science, adopted from physics and even there representing the inappropriate invention of philosophers of science with one-sided views rather than a methodological recipe for success, suggests that our linguistic distinction systems, which precede the collection of data, are in relation to the structure of the world, 'adopted', 'appropriate', 'reflecting', 'model' or otherwise. This is surely correct insofar as even intellectually linguistic classification and hence the making of distinctions, can only take place where there is actually something present which can be classified, distinguished or observed. (Even the production of screws is subject to preconditions lying outside the sphere of human activities.) The decisive factor is, however, that the *appropriateness* of distinction systems, and hence the usefulness of terminologies (even that of psychology), can only be more or less successful relative to the *intentions* which guide the distinctions. The usual view, that one wishes to see the world, the human being or whatever, as they 'really' are – unconditionally and without special interests with regard to this knowledge, is mere theoretical-linguistic thoughtlessness. All distinction systems in everyday and colloquial language as well as those in artificial scientific languages have de facto aims relative to which they are more or less successful. Of course there is again here the *de facto* possibility of not explicitly discussing these aims and criteria of success but this does not alter the fact that distinction systems, too, can be more or less successful and are hence to be seen in relation to aims.

The absurdity of the empirical investigation of the properties of screws has, therefore, the following parallel in the absurdity of a descriptive ecological developmental psychology. As long as it is not explicitly stated with respect to *which* criteria of success, that is to say, to *which* intentions with regard to application or knowledge incorporating the distinctions are made, variables are defined, there is not the slightest guarantee that developmental psychologists will not – metaphorically speaking – correlate the color and length of screws and be made to look absurd by the next batch of screws produced by an automatic screw machine. Setting oneself the aim of arriving at 'true' theories leads nowhere; to declare this aim would be circular.

What appears, therefore, on the one hand, to be critical methodological abstention of descriptivism compared with causalism, that is to say, abstention which dispenses with problematical causal statements is, on the other hand, bound up with renouncing possibilities of recognizing relations of the conceptual, methodologically prior structuring of the areas of environment and development (as must be conceded to the causalist, in the light of his intention of wanting to 'explain' what has been 'observed'). Descriptivism is the final consequence of a naive realistic tendency, stemming from the tradition of empiricism in the sciences, which regards the task of the empirical scientist as being the description of 'existing' structures, and, in so doing, overlooks the constitutive role of specialized terminology in the structuring of the area of investigation. Put more trenchantly, the descriptivist does not know what he is talking about when he makes his correlation assertions and provides empirical evidence for them, since the distinctive intentions of his conceptual system are not descriptively accessible to him.

On (d): *Relativism*

From whatever angle the attention of the researcher into developmental psychology is directed at the environment, for the respective individual being studied, only environmental factors with which the individual has come into contact can be of significance, and this not only in a trivial sense. Even before each empirical investigation, it may be stringently argued that environment can only begin to act upon the individual when subject to subjective susceptibilities and peculiarities of the active, dynamic acquisition of the environment by the individual. That is to say, environment is to be seen relative to individuals and not only in the trivial sense that, from the spectrum of all "possible" (i.e. describable by conceptual structuring on the part of the scientist) components, the ones existing for the respective individual are to be taken into account. It would therefore seem obvious to the psychologist who is critically aware of the traps of descriptivism that it is necessary to make the subjectively manifested "effects" of the environment on the individual the "objective" data. This position may be called *relativistic*.

Closer examination shows, however that the repair work involved in relativizing environmental data to individuals, which is, admittedly, still intended to lead to the collection of data within the framework of an empirical research which does not make its aims explicit, cannot succeed. No objective basis for data can be created. Throughout his whole life each individual is involved in conflict with his environment – physical and social – a conflict which only develops into a mixture of success and failure relative to respective subjective intentions. On the other hand, success and failure do not exclusively depend on these intentions. Whether an individual selects appropriate means to attain his ends is a question which is decided outside of the individual – as is formulated for example, in such expressions as "in accordance with the laws of nature" or as a result of "social pressure". If we examine the question more closely, it is not sufficient to abandon the naive two-world-thinking – environment versus individual – of the causalist and the descriptivist and to see 'environmental data' as relative to individuals and their active and passive selectional processes. The active and passive individual's acts of selection vis-à-vis the environment are only trained under certain conditions so that regarding environmental factors as being relative to individuals is itself dependent on concrete environmental factors. Hence it is clear that we have here not a simple ('relative') but a reflexive relationship between the environment and individuals. The possibility of "collecting data", therefore, becomes an illusion as long as one seeks a general structuring of the totality of environmental and individual reflexively interwoven variables, a structuring directed towards increase in knowledge. Thus, relativism either falls short of the mark if it only takes account of the subjective aspects of environmental data or it must, of necessity, prove inadequate to the task of establishing a definitional foundation which will permit the "objective" collection of data.

2. TOWARD AN ALTERNATIVE PROCEDURE

My criticism of the four empiricist positions with regard to prior decisions necessary for the collection of data represents, in short, an answer to the question as to how one is not able to conduct empirical developmental psychology. Lest my criticism be misunderstood as suggesting that empirical developmental psychology – or ecological psychology – is here being represented as methodologically impossible I shall now attempt to delineate an approach which avoids the preliminary problems in collecting data as outlined above. What most significantly distinguishes this approach from other established approaches is the abandonment of the aim of first launching an empirical investigation and afterwards considering how the results are to be evaluated.

My starting point is the fact that in almost all human societies 'education' takes place. Less factual than normative is my second premise that every educator should be able, at least on being asked, to state to what end he is educating those entrusted to him. At all events, one must refute the objection frequently made to discussions of the aims of education that (quite apart from the problem of responsibility or authority) it is impossible to determine the aims of education, since this implies the reduction of human behavior to norms. The fact that education takes place is evidence that it is at least possible for the educators to choose their aims.

It is probably also true to say that the choice of educational means for the attainment of particular educational ends is occasionally unsuccessful. This might, I think, be a starting point for research into developmental psychology. It would accordingly be one task (amongst others) of developmental psychologists to search for educational knowledge, which is, of course, to be regarded as knowledge concerning the choice of means, not for any but for particular educational aims. For particular educational aims the following procedure can be adopted. Where structures which are no longer naively assumed to be present are sought or investigated even the empirical investigator requires a criterion for distinguishing 'true' from 'false' results (and this criterion cannot be 'correspondence with reality' on account of the structuring of reality by the researcher himself). As such a criterion of distinction I would suggest making hypotheses about (as one can again suggestively say) the "causes" of the unsuccessful choice of means of education by educators, or, more precisely, the truth or falsehood of these hypotheses dependent on their practical success. In other words, if a hypothesis for the explanation of failure in education (only definable relative to educational aims) finally leads to other measures (in the form of assistance offered to educators on the part of a science of education) which help to avoid the uncontroversial failure, then this hypothesis (the system of hypotheses) may be regarded as true. That we are here concerned with empirical truth is clear.

If a developmental psychologist adopts such a criterion of truth (which, in the example quoted, makes developmental psychology an auxiliary science of pedagogics, and above all of practical pedagogics) then the above mentioned problems concerning the formation of concepts, which make the collection of data possible, no longer exist. The formulation of educational aims also, of course, requires talking about the actions of those being educated in certain situations and will, thus, lead to a discussion of individuals and 'environment'. This formation of concepts naturally takes place within the context of discussion of the justification of moral and political norms (for example the norm of educating someone toward intellectual and moral independence).

It would be wrong to think that, in doing this, we are merely transferring the problem of definition of variables in the traditional sense to definitions of variables in the field of discussion of educational aims. No researcher is here confronted with a tabula rasa where, on the one hand, he has the freedom and, on the other hand, the trouble of having to find the most elementary points of orientation. The point of departure is much more the fact 'education', the natural and cultural scope for the actions of the educator. The 'variable' is here (within limits) the choice of educational methods by the educator. In respect of the meta-scientific goal suggested here, namely that of seeking knowledge of developmental psychology, which can then be applied to practice, the starting point must be the factually manageable components of the sphere of investigation. This means it must be the actions and measures of the educators and not the desired behavior of those being educated.

If one now finally asks about conditions for the possibility of an empirical developmental psychology, the conceptual structuring of the reality to be empirically investigated is embedded in contexts of action such as education situations (ranging from the possibilities of action within the framework of personal parent-child relations to the educational conditions of driving instruction for the purpose of acquiring a driving licence) and systems of purposes of educators. Here 'data' can be collected for purposes of 'explanation' in a manner which can claim to be scientific in the sense of trans-subjective testability, since there are criteria for the testing of the truth of the explanatory statements of the developmental psychologist. If his explanation of the failure of educational measures leads to other successful measures, it is true. This gives rise to a situation which is paradoxical for an empiricist understanding of science in that the psychologist conducting empirical research into the individual in his environment from a "scientific" distance (that is to say from norms of actions) must begin with various 'scopes' for action as the only variables in the strict sense of the word and must hence remain oriented towards the goals of the action (in the stated example of the actions of the educator).

The question whether individuals are for the most part determined by their environment or construct their own respective environment, can neither be decided empirically nor be a mere previous methodological decision. Only terminological clarification of the talk about psychological theories or research tasks can at most offer an answer to this question (which offers a critical argument against stimulus-response schemata). The question is otherwise irrelevant for the developmental psychologist.

NOTES ON THE CONTRIBUTORS

KUNO LORENZ, born 1932, studied Mathematics, Physics and Philosophy in Tübingen, Hamburg and Bonn; 1958 Visiting Fellow, Princeton University/USA; 1961 Ph.D. at Kiel University under the supervision of Paul Lorenzen (Dissertation on "Arithmetik und Logik als Spiele"); 1969 Habilitation in Philosophy, Erlangen. 1970–1974 Full Professor of Philosophy at Hamburg University; since 1974 at the 'Universität des Saarlandes' (Saarbrücken). Visiting Professor in Bern (1980), Aix-en-Provence (1983), and Haifa (1984).

Books: Elemente der Sprachkritik. Eine Alternative zum Dogmatismus und Skeptizismus in der Analytischen Philosophie (1970); (with P. Lorenzen) *Dialogische Logik* (1978); (ed.) *Konstruktionen versus Positionen. Beiträge zur Diskussion um die konstruktive Wissenschaftstheorie*, vols. I–II (1978); (ed.) *Richard Gätschenberger: Zeichen. Die Fundamente des Wissens* (1977); (ed.) *Identität und Individuation*, vols. I–II (1982); (ed.) (with M. Dascal and D. Gerhardus and G. Meggle) *Philosophy of Language. An International Handbook of Contemporary Research* (in preparation).

Articles (selection): "Über Strukturverbände von Verbandsgruppen" (1962); "Der dialogische Wahrheitsbegriff" (1972); "Die Monadologie als Entwurf einer Hermeneutik" (1975); "On the Relation between the Partition of a Whole into Parts and the Attribution of Properties to an Object" (1977); "Moralphilosophische Argumentationen bei Camus" (1978); "The Concept of Science. Some Remarks on the Methodological Issue 'Construction' versus 'Description' in the Philosophy of Science" (1979); "Sprachphilosophie" (1980); "Semiotic Stages in the Genesis of Individuals" (1981); "Erleben und Erkennen. Studien der Erkenntnis bei Moritz Schlick" (1982); "Vom Sein und vom Sollen" (1982); "O quadro conceptual da lógîca dialógîca" (1985); Contributions to the *Historisches Wörterbuch der Philosophie* (ed. J. Ritter†, *et alii*) (1971ff.) and to the *Enzyklopädie Philosophie und Wissenschaftstheorie* (ed. J. Mittelstrass) (1980ff.).

CARL FRIEDRICH GETHMANN, born 1944, studied Philosophy in

Robert E. Butts and James Robert Brown (eds.), Constructivism and Science, 271–278.

Bonn, Innsbruck and Bochum; 1968 lic. phil., Institutum Philosophicum Oenipontanum; 1971 Ph.D. at Bochum University; 1978 Habilitation in Philosophy at Constance University. 1968 Assistant Professor, Lecturer of Philosophy (Essen); 1978 Visiting Professor (Constance); 1979 Associate Professor of Philosophy (Essen). Visiting Professor in Düsseldorf, Duisburg and Göttingen.

Books: Verstehen und Auslegung. Untersuchungen zum Methodenproblem in der Philosophie Martin Heideggers (1974); *Protologik. Untersuchungen zur formalen Pragmatik der Normenbegründung* (1979); (ed.) *Theorie des wissenschaftlichen Argumentierens* (1980); *Logik und Pragmatik. Zum philosophischen Rechtfertigungsproblem logischer Kalküle* (1982); (ed.) *Lebenswelt und Wissenschaft. Studien zum Verhältnis von Phänomenologie und Wissenschaftstheorie* (1985).

Articles (selection): "Logische Propädeutik als Fundamentalphilosophie" (1968); "Heideggers These vom Sein des Daseins als Sorge und die Frage nach der Subjektivität des Subjekts (1970); "Zu Heideggers Wahrheitsbegriff" (1974); "Logische Deduktion und transzendentale Konstitution" (1974); "Deontische Logik und Verallgemeinerbarkeit" (1977); "Das Problem der Begründung zwischen Dezisionismus und Fundamentalismus" (1977); "Ist Philosophie als Institution möglich?" (1978); "Zur formalen Pragmatik der Normenbegründung" (1979); "Die Logik der Wissenschaftstheorie" (1980); "Wissenschaftsforschung?" (1981); "Proto-Ethik" (1981); "Zur methodischen Ordnung logischer Kalkültypen" (1982); "Formale Logik und Dialektik. Zur Logik-Diskussion in der DDR" (1984); "Heideggers 'existenzialer Begriff der Wissenschaft'" (1985); "Handlung – Bedeutung – Folgerung. Probleme des methodischen Aufbaus der Logik" (1985).

FRIEDRICH KAMBARTEL, born 1935, studied Chemistry, Physics, Mathematics, Mathematical Logic and Philosophy in Münster; 1959 Dr. rer. nat.; 1966 Habilitation in Philosophy. 1960–1966 Assistant Professor of Philosophy, Münster; since 1966 Full Professor of Philosophy, Constance.

Books: Orthonormale Systeme und Randintegralformeln in der Funktionentheorie mehrerer Veränderlichen (1960); (ed.) *Bernard Bolzanos Grundlegung der Logik. Ausgewählte Paragraphen aus der Wissenschaftslehre*, vols. I–II (1963, ²1978); *Erfahrung und Struktur. Bausteine zu einer Kritik des Empirismus und Formalismus* (1968, ²1976); *Was ist und soll Philosophie* (1968, ²1974); (ed.) *Gottlob Frege: Nach-*

gelassene Schriften (1969, ²1983, Engl. Posthumous Writings (1979)); (ed.) *Praktische Philosophie und Wissenschaftstheorie* (1974); (with P. Janich and J. Mittelstrass) *Wissenschaftstheorie als Wissenschaftskritik* (1974); *Theorie und Begründung. Studien zum Philosophie- und Wissenschaftsverständnis* (1976); (ed.) *Gottlob Frege: Wissenschaftlicher Briefwechsel* (1976).

Articles (selection): "Mathematics and the Concept of Theory" (1965); "Ethik und Mathematik" (1972, Engl. 'Ethics and Mathematics' (1984)); "Wie abhängig ist die Physik von Erfahrung und Geschichte? Zur methodischen Orientierung apriorischer und empirischer Elemente der Naturwissenschaft" (1973); "Bemerkungen zum normativen Fundament der Ökonomie" (1975); "Frege und die axiomatische Methode. Zur Kritik mathematikhistorischer Legitimationsversuche der formalistischen Ideologie" (1975); "Symbolic Acts. Remarks on the Foundations of a Pragmatic Theory of Language" (1976); "Ist rationale Ökonomie als empirisch-quantitative Wissenschaft möglich?" (1978); "Universalität als Lebensform" (1978); "Überlegungen zum pragmatischen und zum argumentativen Fundament der Logik" (1979); "Pragmatic Reconstruction, as Exemplified by an Understanding of Arithmetics" (1980); "Friede und Konfliktbewältigung, handlungstheoretisch und ethisch betrachtet" (1981); "Rekonstruktion und Rationalität. Zur normativen Grundlage einer Theorie der Wissenschaft" (1981); (with H.-J. Schneider) "Constructing a Pragmatic Foundation for Semantics" (1981); "Zur grammatischen Unmöglichkeit einer evolutionstheoretischen Erklärung der humanen Welt" (1984).

JÜRGEN MITTELSTRASS, born 1936, studied Philosophy, Protestant Theology, German Literature in Erlangen, Hamburg and Bonn; 1961–1962 postgraduate study at Oxford University; 1961 Ph.D. in Philosophy (Erlangen University); 1968 Habilitation in Philosophy (Erlangen University). 1962–1970 Assistant Professor of Philosophy at Erlangen. 1970 Visiting Professor at Temple University, Philadelphia. Since 1970 Full Professor of Philosophy at Constance University.

Books: Die Rettung der Phänomene: Ursprung und Geschichte eines antiken Forschungsprinzips (1962); *Neuzeit und Aufklärung: Studien zur Entstehung der neuzeitlichen Wissenschaft und Philosophie* (1970); *Das praktische Fundament der Wissenschaft und die Aufgabe der Philosophie* (1972); (ed., with F. Kambartel) *Zum normativen Fundament der Wissenschaft* (1973); *Die Möglichkeit von Wissenschaft* (1974); (with P. Janich and F. Kambartel) *Wissenschaftstheorie als Wissenschaftskritik* (1974); (ed.) *Methodologische Probleme einer normativ-kritischen Gesellschaftstheorie* (1975); (ed., with M. Riedel) *Vernünftiges Denken: Studien zur praktischen Philosophie und Wissenschaftstheorie* (1978); (ed.) *Methodenprobleme der Wissenschaften vom gesellschaftlichen Handeln* (1979); *Wissenschaft als Lebensform: Reden über philosophische Orientierungen in Wissenschaft und Universität* (1982); *Fortschritt und Eliten: Analysen zur Rationalität der Industriegesellschaft* (1984); (ed.) *Enzyklopädie Philosophie und Wissenschaftstheorie*, vols. I–III (vol. I 1980, vol. II 1984, vol. III in preparation).

Articles (selection): (with K. Lorenz) "On Rational Philosophy of Language: The Programme in Plato's *Cratylus* Reconsidered" (1967); "Predication and Recurrence of the Same" (1968); "The Galilean Revolution: The Historical Fate of a Methodological Insight" (1972); "Methodological Elements in Keplerian Astronomy" (1972); (with E. J. Aiton) "Leibniz: Physics, Logic, Metaphysics" (1973); "Violence and the Limits of Toleration" (1974); "Changing Concepts of the *a priori*" (1977); "'Phaenomena bene fundata': from 'Saving the Appearances' to the Mechanisation of the World-Picture" (1978); "The Philosopher's Conception of *Mathesis Universalis* from Descartes to Leibniz" (1979); "What does 'Reconstruction' Mean in the Analysis of Science and Its History?" (1980); "Towards a Normative Conception of the Growth of Knowledge" (1980); "Substance and Its Concept in Leibniz" (1981); "Scientific Rationality and Its Reconstruction" (1985); "Leibniz and Kant on Mathematical and Philosophical Knowledge" (1985).

PAUL LORENZEN, born 1915, studied Mathematics, Physics and Philosophy in Kiel, Berlin, Göttingen; 1938 Ph.D. Mathematics in Göttingen; 1946 Habilitation. 1952 Professor of Mathematics (Bonn University); 1956 Full Professor of Philosophy (Kiel), 1962–1980 (Erlangen University). Since 1960 member of the Academy of Sciences, Göttingen. Visiting Professor in Cambridge (England), Princeton, Stanford, Austin (Texas), Moscow, Oxford, Boston. 1974 Dr. h.c. from State University Rio de Janeiro, 1985 Dr. h.c. from Boston University.

Books: Einführung in die operative Logik und Mathematik (1955, ^{2}1969); *Formale Logik* (1958, ^{4}1970, Engl. *Formal Logic* 1965, Span.

Lógica formal 1970); *Die Entstehung der exakten Wissenschaften* (1960); *Metamathematik* (1962, ^{2}1980, French *Métamathématique* 1967, Span. *Metamatemática* 1971); *Differential und Integral. Eine konstruktive Einführung in die klassische Analysis* (1965, Engl. *Differential and Integral. A Constructive Introduction to Classical Analysis* 1971); (with W. Kamlah) *Logische Propädeutik oder Vorschule des vernünftigen Redens* (1967, ^{2}1973); *Methodisches Denken* (1968, ^{2}1974); *Normative Logic and Ethics* (1969, ^{2}1984); (with O. Schwemmer) *Konstruktive Logik, Ethik und Wissenschaftstheorie* (1973, ^{2}1975); *Konstruktive Wissenschaftstheorie* (1974); *Theorie der technischen und politischen Vernunft* (1978); (with K. Lorenz) *Dialogische Logik* (1978); *Elementargeometrie. Das Fundament der Analytischen Geometrie* (1984); *Grundbegriffe der technischen und politischen Kultur* (1985).

OSWALD SCHWEMMER, born 1941, studied Philosophy, Theology, Sociology and Political Science in Bonn, Pullach, Munich and Erlangen. 1970 Doctorate; 1975 Habilitation in Philosophy. 1970–1976 Assistant Professor, 1976 Associate Professor (Erlangen); Visiting Professor in Hamburg, Frankfurt, Göttingen and at the School of Technology in Aachen; 1979–1982 director – together with P. Lorenzen – of the Interdisciplinary Institute for Philosophy and History of Science (Erlangen); since 1982 Full Professor of Philosophy (Marburg).

Books: Philosophie der Praxis. Versuch zur Grundlegung einer Lehre vom moralischen Argumentieren in Verbindung mit einer Interpretation der praktischen Philosophie Kants (1971, ^{2}1980); (with P. Lorenzen) *Konstruktive Logik, Ethik und Wissenschaftstheorie* (1973, ^{2}1975); *Theorie der rationalen Erklärung. Zu den methodischen Grundlagen der Kulturwissenschaften* (1976); (ed.) *Vernunft, Handlung und Erfahrung. Über die Grundlagen und Ziele der Wissenschaften* (1981); *Handlung und Struktur. Zur Wissenschaftstheorie der Kulturwissenschaften* (1985).

Articles (selection): "Vom Widerspruch in der Sache selbst. Ein Versuch, Dialektik als Methode der Kritischen Theorie zu verstehen" (1978); "Verstehen als Methode. Vorüberlegungen zu einer Theorie der Handlungsdeutung" (1978); "Die Vernunft der Wissenschaften. Kritische Bemerkungen zu einem unvermeidlichen Anspruch" (1981); "Die Bildung der Vernunft aus der Erfahrung. Zur Grundlegung einer philosophischen Anthropologie" (1982); "Die praktische Ohnmacht der reinen Vernunft. Bemerkungen zum kategorischen Imperativ

Kants" (1983); "Das Faktum der Vernunft und die Realität des Handelns. Kritische Bemerkungen zur transzendentalphilosophischen Normbegründung und ihrer handlungstheoretischen Begriffsgrundlage im Blick auf Kant" (1985); "Der Universalitätsanspruch der Vernunft und der Wandel moralischer Erfahrungen. Philosophische Bemerkungen zum Werte- und Normenwandel" (1985).

CHRISTIAN THIEL, born 1937, studied Philosophy, Mathematics, Sociology, Psychology (Erlangen), and Art at the Academy of Fine Arts (Munich); 1965 Ph.D. at Erlangen University; 1966 Post-doctoral Fellow at the University of Texas, Austin; 1967 Assistant Professor of Philosophy (Austin); 1970 Habilitation in Philosophy (Erlangen). 1970–1972 Associate Professor (Kiel and Constance); 1972 Full Professor of Philosophy and Philosophy of Science at the Technische Hochschule in Aachen; since 1982 Full Professor of Philosophy at Erlangen University and acting director of its Interdisciplinary Institute for Philosophy and History of Science.

Books: Sinn und Bedeutung in der Logik Gottlob Freges (1965, Engl. *Sense and Reference in Frege's Logic*, 1968, Span. 1972); *Grundlagenkrise und Grundlagenstreit. Studie über das normative Fundament der Wissenschaften am Beispiel von Mathematik und Sozialwissenschaft* (1972); (ed.) *Frege und die moderne Grundlagenforschung* (1975); (ed.) *Erkenntnistheoretische Grundlagen der Mathematik* (1982), as well as a centenary edition of Frege's *Die Grundlagen der Arithmetik* (1985), and (co-ed.) Frege's *Wissenschaftlicher Briefwechsel* (1976, 1980).

Articles (selection): "Würfel aus lauter Würfeln" (1969); "Wozu Geschichte der Logik?" (1972); "Gottlob Frege: Die Abstraktion" (1972, ^{2}1979, enlarged ^{3}1985); "Was heißt 'Wissenschaftliche Begriffsbildung'?" (1973); "Grundlagenforschung und Grundlagen der Wissenschaften" (1974); "Zur Inkonsistenz der Fregeschen Mengenlehre" (1975); "Wahrheitswert und Wertverlauf" (1976); "Que significa 'Construcitivismo'?" (1977); "Leopold Löwenheim: Life, Work, and Early Influence" (1977); "Leibnizens Definition der logischen Allgemeingültigkeit und der 'arithmetische Kalkül'" (1980); "Duality Lost? Transforming Gentzen Derivations into Winning Strategies for Dialogue Games" (1978); "Lakatos' Dialektik der mathematischen Vernunft" (1981); "From Leibniz to Frege: Mathematical Logic Between 1679 and 1879" (1982); "Die Revisionsbedürftigkeit der logischen Semantik

Freges" (1983); "Clavius und die Consequentia Mirabilis" (1983); "Folgen der Emigration deutscher und österreichischer Wissenschaftstheoretiker zwischen 1933 und 1945" (1984).

PETER JANICH, born 1942, studied Physics, Mathematics, Psychology and Philosophy from 1961–1967 at the Universities of Erlangen and Hamburg; 1969 Dr. phil. 1969/1970; Visiting Lecturer in Philosophy, University of Texas at Austin; 1970/1971 Assitant Professor, Erlangen; 1971–1980 Associate Professor of Philosophy and History of Natural Science at the University of Constance; since 1980 Full Professor of Philosophy, Marburg. 1987 Fellow at the Center for Philosophy of Science, Pittsburgh.

Books: Die Protophysik der Zeit (1969); (with F. Kambartel and J. Mittelstrass) *Wissenschaftstheorie als Wissenschaftskritik* (1974); *Die Protophysik der Zeit. Konstruktive Begründung und Geschichte der Zeitmessung* (1980, Engl. Protophysics of Time 1985); (ed.) *Wissenschaftstheorie als Wissenschaftsforschung* (1981); (ed.) *Methodische Philosophie. Beiträge zum Begründungsproblem der exakten Wissenschaften in Auseinandersetzung mit Hugo Dingler* (1984). *Euklids Erbe. Ist der Raum dreidimensional?* (1989).

Articles (selection): "Zweck und Methode der Physik aus philosophischer Sicht" (1973); "Galilei und die Galileiinvarianz" (1974); "Zur Protophysik des Raumes" (1976); "Ist 'Masse' ein theoretischer Begriff?" (1977); "Die Sprache der Physik und die Wirklichkeit der Naturwissenschaften" (1977); "Möglichkeiten und Grenzen quantitativer Methoden" (1978); "Physics, natural science or technology?" (1978); "Umweltdeterminiertheit oder Konstruktion der Wirklichkeit?" (1979); "Das Mass der Masse" (1979); "Physikalische Begriffsbildung gegen das Prinzip der methodischen Ordnung?" (1979); "Die erkenntnistheoretischen Quellen Albert Einsteins" (1979); "Newton *ob omni naevo vindicatus*" (1981, Engl. 1982); "Natur und Handlung. Über die methodischen Grundlagen naturwissenschaftlicher Erfahrung" (1981); "Ist Psychologie auf der Grundlage technischer Rationalität als Wissenschaft möglich?" (1981); "Hugo Dingler, die Protophysik und die spezielle Relativitätstheorie" (1984); "Hat Ernst Mach die Protophysik der Zeit kritisiert?" (1985); "Die Eindeutigkeit der Massenmessung und die Definition der Trägheit" (1985); "Die Galileische Geometrie. Zum Verhältnis der geometrischen Idealisierung bei E.

Husserl und der protophysikalischen Ideationstheorie" (1985); "Technik in der Naturwissenschaft und Naturwissenschaft in der Technik" (1985); "L'operationalismo come criterio fondamentale di scientificità" (1985). "Evolution der Erkenntnis oder Erkenntnis der Evolution?" (1987); "Geschwindigkeit und Zeit. Aristoteles und Augustinus als Lehrmeister der modernen Physik?" (1988); "Physiology and Language. Epistemological Questions about Scientific Theories of Perception" (1989).

INDEX

(Compiled by Philip Catton)

a priori
as *Begründung* of theory, xix, 83, 87
distinction-, 91, 93, 97, 102n
foundational versus proof-theoretic, 87
graded (according to Mittelstrass' scheme), xvii, xviii, xix, xxii, 90–98
measuring-, 90, 138
pragmatic, xvi, xvii, xviii, xix, 92–98
pre-theoretical, 90
probabilities, 163
synthetic, 84–87
technical-, 100n
the *Lebenswelt* and, xvii, 80, 87–91
transcendental, 87, 88
truth, 15, 86, 87
action, purposive *See* practical action
Albert, H., 185
Almeder, R., 236n
Angelelli, I., 99n
antinomy
idea explicated, 24
of freedom, 83
See also paradox
Apel, K.-O.
defends is-ought gap 18n
in "transcendental-philosophic" as opposed to "skeptical decisionistic" tradition, 185
mentioned, 101n
Archimedes, 133
Aristotle
his dynamics, 153
on intuitionistic mathematics, xxii
on logic — art or science?, 62–63, 64, 74n, 75n
on means and ends, 18n
on reasons and causes, 17n
on the good life, 173, 175, 243
Aschenberg, R., 98n
Austin, J. L., 48

Baum, M., 98n
Becker, O., 129
Begründung concept
in constructivist reinterpretation of rationality of science, xix
mentioned, 18n
the a priori and, 83, 87
Beitkopf, A., 26
Bennett, J., 98n
Berge, C., 75n
Berger, P. L., 218n
Bergström, L., 17n
Berk, U., 57n
Berkeley, G., 108
Bernays, P., 124, 126n
Bernoulli, J., 167, 168
Blaschke, W., 138
Böhme, G., xiii
Bollhagen, P., 238n
Borel, E., 167, 169
Bourbaki group, 128
Bridgman, P., 146
Brouwer, L.
his intuitionistic mathematics, xxii, 21, 22, 73n, 107
influence on constructivism, xxii
influence on philosophy of science, xxii, 21, 22
on *tertium non datur*, 59
operative logic and, 75n
presented no full formal system of logic, 61
recognized that classical logic applies to finite mathematics, 62, 74n
Bubner, R.
contends Kant's *MFNS* is protophysics, xxivn

mentioned, xxiiin, 120n
Buchdahl, G., 99n
Buridan, J., 64
Butts, R. E., 101n

Cantor, G., 106, 110
Carnap, R.
empiricist, 7, 146
Frege and, 48, 108
in Germanic tradition, xiii
logicist about mathematics, 108
mentioned, 101n, 151
non-dogmatic about competing logics, 21–23
on correspondence rules, xv
on dispositional predicates, 26–27
operationalism and, 149
principle of tolerance, 22–3, 29
rational reconstruction and, 94, 146
Cartan, H., 169
Cassirer, E.: on pragmatic a priori, xviii
Cauchy, A. L., 106
concepts, theoretical: creation modelled on human purposive action, xvi
conventionalism
Carnap's principle of tolerance and, 22–23, 29–30
constructivism and, xx, xxivn
dialogic and, xx
logic and, 21, 22–23, 29–30
operationalism and, xx, xxivn, 260–261
realism and, xx
Coulomb, C. A., 140
Crawford, P. A., 98n

Dahrendorf, R., 236n
Davidson, D., 17n
dialogic
compared to classical logic, xxi, 72–73
"dialogue-definiteness", 66, 69, 72
discussed in some depth, 66–73
element of arbitrariness in, 55
operational character of constructivism and, xxi, 60
pragmatic context for logical truth and, xxi, 17n, 18n, 54, 62–66
Dedekind, R., 121
Descartes, R., 127
Dingler, H.
as Kantian, xiii, xiv
"father" of constructivism, xiii
mentioned, 108

Ebbinghaus, K., 74n
Einstein, A.
his dynamics and protogeometry, 138–144
mentioned, xiii
realism and, xvi
empiricism
about mathematics and logic, 61, 62
about meaning, xv
about probability, 163
analyticity and, 87
as response to scepticism, 4
contrasted with constructivism, xvi
empirical versus transcendental idealism, 80
"evolutionarism" and, 4, 6
Kantian *empirische Erkenntnisse* and, 88–89, 90
measurement and, 86–87
of Carnap and Stegmüller, 7
theoretical terms and, 145
versus rationalism, 4, 9
Engels, F., 236n
enlightenment, 56, 186
Erlangen school, xiv, 7, 14, 41n
Erlanger Programm, 127
Esser, A., 237n
Essler, W. K., 162n
Euclid, 105, 110, 125, 127–138

Feyerabend, P.
historicism and, 7
versus Sneed, 146
finalization, xiii, 176–178, 202
finitism, 21–22
foundationalism
contrast (of positivist variety) with constructivism, xiv
infinite regress problem for, xvii
Kant's view of, 83
pragmatics not foundationalist, 194

"science as presentation" and, 7
transcendental a priori and, 87
foundations of mathematics, xxii, xxiii, 93, 105–126
Gödel result and, 112
practical action and, 116–118
foundations of science
a concern of *Wissenschaftstheorie*, xiii
as methodical constructions, 83
interpreted by constructivism, xiv–xix, xxii
Kant's view of, 83
the concern of protoscience, xxii, xxiii
transcendental a priori and, 87
See also Begründung concept
Frankfurt school: in relation to Hegel and Marx, xiii
Fraenkel, A., 111
Fréchet, M., 167
freedom
eighteenth-century fanfare about, 173
justice and, 11, 12
Frege, G., 19, 21, 106–108, 111, 114

Galileo Galilei, 153, 162n, 260
Gauss, K. F., 127
Gätschenberger, R., 17n
Gentzen, G., 42n
geometry
pragmatic a priori and, xviii
priority to physical theory, 127
protoscience and, xxii, 14–15, 129
structuralism and, 127–128
"Thaletic", 129–137
Gerhardus, D., 16n
Gethmann, C. F., xxi, xxiii, 42n, 57n, 102n, 218n
Gödel, K., 74n, 112, 113
Goodman, N., 40, 42n
Gram, M. S., 98n
Griffiths, A. P., 98n

Habermas, J., xiii, 101n
in "transcendental-philosophic" as opposed to "skeptical decisionistic" tradition, 185
Hegselmann, R., 40, 42n, 102n, 218n
Hahn, H., 21, 30
Heidegger, M., xiii
Hegel, G., xiii, 173, 221, 222, 236
Heinrich, D., 102n
Heinsoeth, H., 99n
Held, V., 236n
Helmholtz, H. v., 128, 148, 149, 162n
Hempel, C. G., 18n, 24–25, 145, 146, 149, 151, 161n
hermeneutical philosophy: in relation to the tradition of phenomenology, xiii
Hermes, H., 16n
Heyting, A., 22, 56n, 60, 74n
Hilbert, D., 21, 107, 111, 112, 113, 114, 123, 124, 127, 128, 129, 132, 133, 136, 143
Hinske, N., 99n
Hinst, R., 42n
Hintikka, J., 98n, 99n, 100n
Hirsch-Weber, W., 237n
historicism: of Kuhn and Lakatos, 7
Hobbes, T., 174, 175
Hoffmeister, J., 237n
Hudson, W. D., 18n
Husserl, E.
axiomatic understanding of mathematical theories and, 114
constructivists' use of the *Lebenswelt* concept and, xvi, xvii, xxiiin
mentioned, xiii
Huygens, C., 154
hypothetico-deductivism
realism and, xvi
relation to pragmatic a priori, xix

imperative
afinal and final, 176–180
categorical, 181, 183, 207, 208, 210–211
general final imperatives are maxims, 181, 205
'interest' defined by means of, 225
logification, 213
of reconstruction, 98
universal, 180–181
universal general final imperatives are norms, 181, 205

intentional behavior *See* practical action
intuitionism
  constructivism and, xxii, 123—126
  emphasis on the nonlinguistic, xxii
  foundations of mathematics and, 114, 123—126
  logic of philosophy of science and, 19—20, 21, 31
intuitionistic logic, xxi, 19—20, 21, 24—29, 31, 36, 37, 40, 41, 56n, 59—76 *passim*
is-ought gap, 13, 14, 18n

James, W., 16n
Janich, P.
  his monograph on protophysics recommended, xxivn
  mentioned, xxii, xxiiin, 100n, 162n, 238n, 239n
  on interdependence of constructive and normative approaches to science, 18n
Johansson, I., 42n, 219n

Kambartel, F., xxi, xxiii, 18n, 42n, 56n, 57n, 99n, 100n, 101n, 114, 126n, 218n, 224, 237n, 238n, 239n
Kamlah, W.
  cofounder of "Erlangen School", xiv
  mentioned, 42n, 56n, 100n
Kant, I
  concepts without intuitions are empty, intuitions without concepts are blind, 15
  "grandfather" of constructivism, xiii
  his presuming that there is knowledge and his efforts to clarify conditions of its possibility, xiv, xvii, 16n, 106, 241
  in relation to intuitionistic mathematics, xxii
  on 'interest', 221
  on maxims and norms, 181, 207, 208, 210—211
  on practical reason, 174, 175, 181, 183, 185, 186, 207, 208, 210—211, 218n, 221
  on 'transcendental', 77—98 *passim*
  propounds "justification" rather than "grounding" conditions, xix, 6—7, 81, 83, 87, 94
  set on his feet, 91
  whether his *Metaphysical Foundations of Natural Science* is constructivist protophysics, xxivn, 243
  See also transcendentalism
Kapp, E., 75n
Kaufmann, F., 21—22
Kepler, J., 158
Kleene, S. C., 74n
Klein, F., 127
Kolmogorov, A.
  inventor of proof-theoretic interpretation of intuitionistic logic, 60, 74n
  mentioned, 56n
  on "probability fields", 167, 169
  "operative" logic and, 75n
Körner, S., 79, 98n, 102n
Kuhn, T. S.: historicism and, 7
Kutschera, F. v., 26, 219n

Lakatos, I.: rationalism and, 7
Lambert, J. H., 100n
Lang, K., 237n
language
  distinction between formal-analytic/synthetic and material-analytic/synthetic, 87, 88, 93
  importance of what is nonlinguistic, xxii, xxiiin, 47—56, 64—65, 78, 98
  linguistic norms, 11—12, 88, 90
  observational, 147
  ordinary language and language of science to be reconstructed from a fictitious elementary language concerned with "preactions", 8—9
  predication linguistically fundamental, 91
  whether ordinary language about ordinary objects can be extended into scientific domain, 15
Laplace, P.-S. de, 141, 167, 169

Lebenswelt (life-world)
  economics and, 244
  mentioned, 194, 195
  philosophy of language and, xxiiin
  pragmatic a priori and, xvi, xvii, xviii, xxiiin, 80, 90–91
  "primary praxis" and, 65
  protologic and, 30–31, 32, 33–38 *passim*
  protoscience and, xxii
Lebesgue, H., 167, 168, 169
Leibniz, G. W. v.
  conception of logic, 61, 73
  his *ars iudicandi* — *ars inveniendi*, Analysis — Synthesis distinction, 5, 16n
Lenk, H., 57n
Lenzen, W., 24–26
Lewis, C. I., 61
logic
  art or science?, 62–63, 64, 74n, 75n
  Carnap's principle of tolerance and, 22–23, 29–30
  competing varieties, 19–20
  conventionalism and, 22–23, 29–30
  dogmatism about, 20–23, 24, 27, 32, 38–39, 41
  in philosophies of Wittgenstein, 21, 30–31, 77
  in understanding language, xxi
  is specifically human, 38, 61
  justification of, 31, 33–34
  logical and pragmatic reconstruction, 94–95
  many-valued, 59–76 *passim*
  minimal, 27–29
  of practical justification, 175
  paradoxes in philosophy of science and, 24–29, 40–41
  philosophy of science and, 19–46 *passim*
  possible worlds and, 61
  pragmatism and, xxi, 17n, 18n, 29, 31
  "relatively universal", 62
  transcendental, 77
  See also intuitionistic logic
Lorentz, H. A., 141, 142
Lorenz, K.
  attempts to mediate between outlooks of constructivism and classical logic, xxi–xx
  coinventor of dialogic, xxi, 17n, 18n, 54, 55, 124
  formalizes is-ought gap, 18n
  mentioned, xxii, 16n, 57n, 75n, 100n, 102n, 126n
Lorenzen, P.
  cofounder of "Erlangen School", xiv
  coinventor of dialogic, xxi, 17n, 54, 55, 124
  constructivism and, xiv
  defined 'mass', 150
  inventor of operative interpretation of intuitionistic logic, 60, 75n
  mentioned, xvii, xxii, 28, 32, 56n, 57n, 74n, 75n, 100n, 101n, 126n, 162n, 238n
  methodical reconstructions of concepts and, xv
Lübbe, H., 185, 218n
Luckmann, T., 218n
Luhmann, N., 101n
Lukács, G., 230, 238n
Łukasiewicz, J., 74n

Mach, E.: mentioned, xiii, xv
MacIver, R. M., 237n
Marx, K., xiii, 222, 229, 230, 236n, 238n, 239n, 249
mass concept: mentioned, xviii, xxii
Maxwell, J. C., 140–142
maxim
  defined as general final imperatives, 181, 205
  reconstruction of their historical development, 188
  situation of one conflicting with another, 186–190
Mead, G. H., 17n
measurement
  basis in pragmatic a priori, xviii, xxii
  empiricism and, 86–87

geometry and, 127—144 *passim*
measuring a priori, 90, 138
Mill, J. S.
against theoretical terms, xv
meaning empiricism and, xv
Minkowski, H.: "Newton—Minkowskian program", 142—144
Mittelstaedt, P., 161n
Mittelstrass, J.
contends that rationality of science concerns conditions of "grounding" rather than "justification", xix
in relation to Kant, xvii, xix, 16n
mentioned, xxii, xxiii, xxiiin, 56n, 99n, 100n, 101n, 238n, 239n
on graded a priori, xvii, xviii, xix, xxii
on interdependence of constructive and normative approaches to science, 18n
on pragmatic a priori, xvi, xvii, xviii, xix
Montague, P., 42n

Neuendorf, H., 236n
Neurath, O., xiv, 21
Newton, I.
definition of mass, 146—147, 159—161, 162n
his idea of applying terrestrial ballistics (Galileo) to the heavens (Kepler), 158
his mechanics and protogeometry, 138—144
the "Newton—Minkowskian program", 142—144
whether his first law of motion is factual, xx
norms
categorical versus hypothetical, 3
concerning randomness in the theory of probability, 163—164, 169—170
defined as universal general final imperatives, 181, 205
functional, 156
legitimation demands, 174
material-synthetic propositions and, 88
meta-rational, 3
of language, 11—12, 88, 90
operational definition and, 149—150, 156
philosophy of language and, xxiiin, 90
practical activity and, xvii, xviii, xxiiin, 90
reconstruction of their historical development, 188
science and, 18n
technical, practical, and moral, 12
when maxims conflict, 186—190
See also rules, is-ought gap

Offe, C., 236n
operationalism
constructivism and, xix, xx
conventionalism and, xx, xxivn, 260—261
operational definition and norms, 149—150, 156
operative logic, xxi, 60, 75n
R. Carnap and, 149

Padoa's proposition, 153
paradox
Goodman's, 40
idea explicated, 24
in philosophy of science, 24—29
logic and, 24—29, 40—41
of derived obligation, 27—29, 40
of dispositional predicators, 26—27
raven, 24—26, 40
See also antinomy
Pasch, M., 127, 132
Passmore, J., 98n
Peano, G., 113, 121
Peirce, C. S.
his theory of categories, 17n
likened to logical positivists, 6, 16n
on consensus in the limit, 11
William James and, 16n
phenomenology
hermeneutics and, xiii
tradition of, xiii
Piaget, J., 17n
Plamenatz, J. P., 237n

Plato, 64, 173, 175
Poincaré, H., 141
Popper, K.
  on theory-observation distinction, xvi
  rationalism and, 7
  realism and, xvi
practical action
  a primary concern of constructivists, xvi—xxii *passim*
  directions for, 197—201
  distinguished as meaningful units, 176
  foundations of mathematics and, 116—118
  ordinary language and language of science to be reconstructed from a fictitious elementary language concerned with "preactions", 7
  symbolic, 49—56, 56n
  the rationality of science and, xix, 3
practical knowledge (know-how), xiv, xvi—xxii *passim*, 8, 19, 31, 32 *See also* pragmatics
pragmatics
  formal, 196
  justification of logic and, 29, 31
  justification of practical principles of justification, 175, 185—186
  methodologically prior to axiomatics, 37
  of scientific argumentation, 39—40, 47—56 *passim*
  pragmatic considerations in Kant restricted in scope, 82
  "pragmatic self-refutation", 78
  protologic and, 32, 33—38 *passim*
  reconstructions of science and, 95
  term appropriated, 194
  transcendental, 94, 95
pragmatism
  linguistic (*see also* pragmatics), 48
  logical truth and, xxi, 17n, 18n
  radical empiricism and, 16n
  teleology and, 55—56
  the rational life and, 55
  transcendental arguments, methods, or conclusions and, 77—98 *passim*
  truth as consensus and, 11
  versus realism, xxiiin
presuppositions
  concerning form of logic, 30
  investigated by constructivism, xvii, xviii, 77—78
  of scientists' use of language, 30
  of traditional philosophy of science, 4—5
  transcendental arguments (from what is "pragmatically fundamental") and, 77—78
Prior, A. N., 27, 28
probability, 163—172
  empiricism versus apriorism about, 163
  "subjective" and "objective", 163
Proclus, 110
protogeometry, 129, 132
protologic, 28, 32, 33—38 *passim*
protoscience
  idea explicated, xxii, xxiii
  mentioned, 7, 149, 158, 243, 244
  normative character, 9
  pragmatic a priori and, xxii
  why at odds with present-day Anglo-American philosophy of science, xxivn
Putnam, H.
  on theory-observation distinction, xvi
  realism and, xvi
Pythagoras, 131, 133, 136, 137

Quine, W. V.
  as pragmatist pole in Anglo-American philosophy of science, xxiiin
  his empiricism about mathematics and logic, 61, 74n
  mentioned, 75n, 126n
  on grades of theoreticity, 18n
  on semantic ascent, 17n
  on theory assessment, 29
  on theory-observation distinction, xvi
  realism and, xvi

rationalism
  as response to scepticism, 4
  of Popper and Lakatos, 7
  versus empiricism, 4, 9

rationality
inherent conservativism of, 3
logic and, 32
means-end, 12
"meta-competence" and, 3, 4
of science, radically reconceived by constructivism (as concerning conditions of "grounding" rather than "justification"), xix
practical, 173–255 *passim*
praxis and, 11
truth and, 3, 11
Raub, W., 40, 42n
Rawls, J., 11, 17n
realism
constrasted with constructivism, xv, xvi, xx
empiricism and, xv, xvi
Reichenbach, H.
his distinction between contexts of discovery and justification, 5, 16n
mentioned, xiii, xv, 101n
on logical analysis of science, 94
Rescher, N., 235n
Richter, V., 75n
Riemann, G. F. B., 127, 142
Rorty, R. M., 17n, 98n, 102n
Rosenberg, J. F., 98n
Ruf, H. L., 98n
rules
a kind of demand, 201
as imperatives that are not and cannot be complied with always and everywhere, 178–179
context- and party-variant rhetorical rules, versus thematized situation-invariant logical rules, 195–196
for the application of predicates, xv
ideative, 88
Kant's schematism of the pure concepts of the understanding and, 92–93
leading to basic theorems of arithmetic, 118–120
making symbolic action possible, 49
methodological, 38
of dialogic, 54, 66–73
of logic, 21, 36, 38
practical activity and, xvii, xviii, xxiiin, 19, 116–119
See also norms
Russell, B., 16n, 19, 21, 106–108, 111

Searle, J. R.
his attacking the is-ought gap, 18n
his linguistic pragmatism, 48
scepticism, 4
Schafer, W., xxiiin
Schaper, E., 99n
Scharpf, W., 236n
Schlick, M., xiii
Schneider, H., 56n, 57n
Scholz, H., 74n, 126n
Schröder-Heister, P., 42n
Schwemmer, O.
mentioned, xxiii, 56n, 100n, 190n, 218n, 238n, 239n
on interdependence of constructive and normative approaches to science, 18n
reconstruction of Kant's categorical imperative, 207
Sellars, W.
as realist pole in Anglo-American philosophy of science, xxiiin
on theory-observation distinction, xvi
realism and, xvi
Skolem, T., 73n, 112, 126n
Small, A. W., 237n
Sneed, J., 17n, 145, 146, 151, 161n
sociology of science, 39
Socrates, 173, 175
Spinoza, B., 105
Stalnaker, R. C., 42n
Stegmüller, W.
assumes correct the intuitions of special scientists, 31, 38
contends that logic is a theory of argumentation, 33
"dogmatist" about classical logic, 20, 22, 24, 27
empiricist, 7, 17n
formalizer of philosophy of science, xiii, 20, 94

holds "superaffirmative" view of science, 151–152
mentioned, 57n, 98n, 99n, 101n, 162n
regards philosophy of science as applied logic, 38, 94
sharply distinguishes cognitive and social structures of science, 43n
states essential intention of Kant's analyses is methodological, "metatheory of empirical knowledge", 79
Strawson, P. F., 9, 10, 17n, 75n, 80, 99n

Tarski, A., 17n, 47, 48, 75n
Tetens, H., 42n
Thales, 127, 129–138
Thiel, C., xxii
Thüring, B., 150, 162n
Toulmin, S., 41
Tranøy, K. E., 11, 17n
transcendental philosophy, xxivn, 4, 5, 6–7, 77–98 *passim*
"analytical", 80
versus "skeptical decisionistic" tradition, 185
Truman, D. B., 236n
truth
a priori, 15, 86
as consensus, 11
as correspondence, 11, 267
as provability, 60, 86
dialogical definition of logical truth, xxi, 67
"object-competence" and, 4
rationality and, 3, 11
value-definiteness, 36, 59–76 *passim*

Ulrich, F., 237n

van Fraassen, B.: for empiricism, against meaning empiricism, xv
Vienna Circle, 21, 22, 30, 41n, 42n
von Mises, R., 167

Weber, M., 181
Waismann, F., 29, 30–31
Weinberg, S., 144
Weyl, H., 107
Whitehead, A. N., 17n
Winterbourne, A. T., 101n
Wissenschaftstheorie
Anglo-American philosophy of science and, xxiiin
constructivism as, xiii, xix, 7
normative character, 9, 18n
Wittgenstein, L.
ideas about logic in *Philosophical Investigations*, 30–31
"logic is transcendental", 77
on *afinale Aufforderungen*, e.g. slab!, 197
on foundations of mathematics as philosophy, 105, 126n
"operative" logic and, 75n
picture theory as theory of predication, 91
Tractarian "dogmatism" about logic, 21
Wohlrapp, H.
his "three stages" of analytic philosophy of science, 7
mentioned, 9, 16n
Wolters, G., 99–100n
Wright, G. H. v., 99n

Zermelo, E., 110, 111

THE UNIVERSITY OF WESTERN ONTARIO
SERIES IN PHILOSOPHY OF SCIENCE

A Series of Books in Philosophy of Science, Methodology, Epistemology, Logic, History of Science, and Related Fields

1. J. Leach, R. Butts, and G. Pearche (eds.), *Science, Decision and Value*. 1973, vii + 219 pp.
2. C. A. Hooker (ed.), *Contemporary Research in the Foundations and Philosophy of Quantum Theory*. 1973, xx + 385 pp.
3. J. Bub, *The Interpretation of Quantum Mechanics*. 1974, ix + 155 pp.
4. D. Hockney, W. Harper, and B. Freed (eds.), *Contemporary Research in Philosophical Logic and Linguistic Semantics*. 1975, vii + 332 pp.
5. C. A. Hooker (ed.), *The Logico-Algebraic Approach to Quantum Mechanics*. 1975, xv + 607 pp.
6. W. L. Harper and C. A. Hooker (eds.), *Foundations of Probability Theory, Statistical Inference, and Statistical Theories of Science*. 3 Volumes. Vol. I: *Foundations and Philosophy of Epistemic Applications of Probability Theory*. 1976, xi + 308 pp. Vol. II: *Foundations and Philosophy of Statistical Inference*. 1976, xi + 455 pp. Vol. III: *Foundations and Philosophy of Statistical Theories in the Physical Sciences*. 1976, xii + 241 pp.
7. C. A. Hooker (ed.), *Physical Theory as Logico-Operational Structure*. 1979, xvii + 334 pp.
8. J. M. Nicholas (ed.), *Images, Perception, and Knowledge*. 1977, ix + 309 pp.
9. R. E. Butts and J. Hintikka (eds.), *Logic, Foundations of Mathematics, and Computability Theory*. 1977, x + 406 pp.
10. R. E. Butts and J. Hintikka (eds.), *Foundational Problems in the Special Sciences*. 1977, x + 427 pp.
11. R. E. Butts and J. Hintikka (eds.), *Basic Problems in Methodology and Linguistics*. 1977, x + 321 pp.
12. R. E. Butts and J. Hintikka (eds.), *Historical and Philosophical Dimensions of Logic, Methodology and Philosophy of Science*. 1977, x + 336 pp.
13. C. A. Hooker (ed.), *Foundations and Applications of Decision Theory*. 2 volumes. Vol. I: *Theoretical Foundations*. 1978, xxiii + 442 pp. Vol. II: *Epistemic and Social Applications*. 1978, xxiii + 206 pp.
14. R. E. Butts and J. C. Pitt (eds.), *New Perspectives on Galileo*. 1978, xvi + 262 pp.
15. W. L. Harper, R. Stalmaker, and G. Pearce (eds.), *Ifs. Conditionals, Belief, Decision, Chance, and Time*. 1980, ix + 345 pp.

16. J. C. Pitt (ed.), *Philosophy in Economics*. 1981, vii + 210 pp.
17. Michael Ruse, *Is Science Sexist?* 1981, xix + 299 pp.
18. Nicholas Rescher, *Leibniz's Metaphysics of Nature*. 1981, xiv + 126 pp.
19. Larry Laudan, *Science and Hypothesis*. 1981, x + 258 pp.
20. William R. Shea, *Nature Mathematized*. 1983, xiii + 325 pp.
21. Michael Ruse, *Nature Animated*. 1983, xiv + 266 pp.
22. William R. Shea (ed.), *Otto Hahn and the Rise of Nuclear Physics*. 1983, x + 252 pp.
23. H. F. Cohen, *Quantifying Music*. 1984, xvii + 308 pp.
24. Robert E. Butts, *Kant and the Double Government Methodology*. 1984, xvi + 339 pp.
25. James Robert Brown (ed.), *Scientific Rationality: The Sociological Turn*. 1984, xiii + 330 pp.
26. Fred Wilson, *Explanation, Causation and Deduction*. 1985, xviii + 385 pp.
27. Joseph C. Pitt (ed.), *Change and Progress in Modern Science*. 1985, viii + 398 pp.
28. Henry B. Hollinger and Michael John Zenzen, *The Nature of Irreversibility*. 1985, xi + 340 pp.
29. Kathleen Okruhlik and James Robert Brown (eds.), *The Natural Philosophy of Leibniz*. 1985, viii + 342 pp.
30. Graham Oddie, *Likeness to Truth*. 1986, xv + 218 pp.
31. Fred Wilson, *Laws and Other Worlds*. 1986, xv + 328 pp.
32. John Earman, *A Primer on Determinism*. 1986, xiv + 273 pp.
33. Robert E. Butts (ed.), *Kant's Philosophy of Physical Science*. 1986, xii + 363 pp.
34. Ian B. MacNeill and Gary J. Umphrey (eds.), *Advances in the Statistical Sciences*, Vol. I, *Applied Probability, Stochastic Processes, and Sampling Theory*. 1987, xxv + 329 pp.
35. Ian B. MacNeill and Gary J. Umphrey (eds.), *Advances in the Statistical Sciences*, Vol. II, *Foundations of Statistical Inference*. 1987, xvii + 287 pp.
36. Ian B. MacNeill and Gary J. Umphrey (eds.), *Advances in the Statistical Sciences*, Vol. III, *Time Series and Econometric Modelling*. 1987, xix + 394 pp.
37. Ian B. MacNeill and Gary J. Umphrey (eds.), *Advances in the Statistical Sciences*, Vol. IV, *Stochastic Hydrology*. 1987, xv + 225 pp.
38. Ian B. MacNeill and Gary J. Umphrey (eds.), *Advances in the Statistical Sciences*, Vol. V, *Biostatistics*. 1987, xvi + 283 pp.
39. Ian B. MacNeill and Gary J. Umphrey (eds.), *Advances in the Statistical Sciences*, Vol. VI, *Actuarial Science*. 1987, xvi + 250 pp.
40. Nicholas Rescher, *Scientific Realism*. 1987, xiii + 169 pp.
41. Brian Skyrms and William L. Harper (eds.), *Causation, Chance, and Credence*. 1988, xii + 284 pp.
42. William L. Harper and Brian Skyrms (eds.), *Causation in Decision, Belief Change and Statistics*. 1988, xix + 252 pp.
43. R. S. Woolhouse (ed.), *Metaphysics and Philosophy of Science in the Seventeenth and Eighteenth Centuries*. 1988, x + 362 pp.

GPSR Compliance

The European Union's (EU) General Product Safety Regulation (GPSR) is a set of rules that requires consumer products to be safe and our obligations to ensure this.

If you have any concerns about our products, you can contact us on ProductSafety@springernature.com

In case Publisher is established outside the EU, the EU authorized representative is:

Springer Nature Customer Service Center GmbH
Europaplatz 3
69115 Heidelberg, Germany

Batch number: 10755284

Printed by Printforce, the Netherlands